Soft Computing for Control of Non-Linear Dynamical Systems

Studies in Fuzziness and Soft Computing

Editor-in-chief
Prof. Janusz Kacprzyk
Systems Research Institute
Polish Academy of Sciences
ul. Newelska 6
01-447 Warsaw, Poland
E-mail: kacprzyk@ibspan.waw.pl
http://www.springer.de/cgi-bin/search_book.pl?series=2941

Further volumes of this series can be found at our homepage.

Vol. 42. S. Pal, G. Ashish and M. Kundu (Eds.)
Soft Computing for Image Processing, 2000
ISBN 3-7908-1217-X

Vol. 43. L.C. Jain, B. Lazzerini and U. Halici (Eds.)
Innovations in ART Neural Networks, 2000
ISBN 3-7908-1270-6

Vol. 44. J. Aracil and F. Gordillo (Eds.)
Stability Issues in Fuzzy Control, 2000
ISBN 3-7908-1277-3

Vol. 45. N. Kasabov (Ed.)
Future Directions for Intelligent Information Systems on Information Sciences, 2000
ISBN 3-7908-1276-5

Vol. 46. J.N. Mordeson and P.S. Nair
Fuzzy Graphs and Fuzzy Hypergraphs, 2000
ISBN 3-7908-1286-2

Vol. 47. E. Czogała† and J. Łęski
Fuzzy and Neuro-Fuzzy Intelligent Systems, 2000
ISBN 3-7908-1289-7

Vol. 48. M. Sakawa
Large Scale Interactive Fuzzy Multiobjective Programming, 2000
ISBN 3-7908-1293-5

Vol. 49. L.I. Kuncheva
Fuzzy Classifier Design, 2000
ISBN 3-7908-1298-6

Vol. 50. F. Crestani and G. Pasi (Eds.)
Soft Computing in Information Retrieval, 2000
ISBN 3-7908-1299-4

Vol. 51. J. Fodor, B. De Baets and P. Perny (Eds.)
Preferences and Decisions under Incomplete Knowledge, 2000
ISBN 3-7908-1303-6

Vol. 52. E.E. Kerre and M. Nachtegael (Eds.)
Fuzzy Techniques in Image Processing, 2000
ISBN 3-7908-1304-4

Vol. 53. G. Bordogna and G. Pasi (Eds.)
Recent Issues on Fuzzy Databases, 2000
ISBN 3-7908-1319-2

Vol. 54. P. Sinčák and J. Vaščák (Eds.)
Quo Vadis Computational Intelligence?, 2000
ISBN 3-7908-1324-9

Vol. 55. J.N. Mordeson, D.S. Malik and S.-C. Cheng
Fuzzy Mathematics in Medicine, 2000
ISBN 3-7908-1325-7

Vol. 56. L. Polkowski, S. Tsumoto and T.Y. Lin (Eds.)
Rough Set Methods and Applications, 2000
ISBN 3-7908-1328-1

Vol. 57. V. Novák and I. Perfilieva (Eds.)
Discovering the World with Fuzzy Logic, 2001
ISBN 3-7908-1330-3

Vol. 58. D.S. Malik and J.N. Mordeson
Fuzzy Discrete Structures, 2000
ISBN 3-7908-1335-4

Vol. 59. T. Furuhashi, S. Tano and H.-A. Jacobsen (Eds.)
Deep Fusion of Computational and Symbolic Processing, 2001
ISBN 3-7908-1339-7

Vol. 60. K.J. Cios (Ed.)
Medical Data Mining and Knowledge Discovery, 2001
ISBN 3-7908-1340-0

Vol. 61. D. Driankov and A. Saffiotti (Eds.)
Fuzzy Logic Techniques for Autonomous Vehicle Navigation, 2001
ISBN 3-7908-1341-9

Vol. 62. N. Baba and L.C. Jain (Eds.)
Computational Intelligence in Games, 2001
ISBN 3-7908-1348-6

Oscar Castillo · Patricia Melin

Soft Computing for Control of Non-Linear Dynamical Systems

With 112 Figures
and 13 Tables

Physica-Verlag
A Springer-Verlag Company

Prof. Dr. Oscar Castillo
Prof. Dr. Patricia Melin
Tijuana Institute of Technology
Department of Computer Science
P.O. Box 4207
Chula Vista, CA 91909
USA
ocastillo@tectijuana.mx
pmelin@tectijuana.mx

ISSN 1434-9922

Cataloging-in-Publication Data applied for
Die Deutsche Bibliothek – CIP-Einheitsaufnahme
Castillo, Oscar: Soft computing for control of non-linear dynamical systems; with 13 tables / Oscar Castillo; Patricia Melin. – Heidelberg; New York: Physica-Verl., 2001
(Studies in fuzziness and soft computing; Vol. 63)
ISBN 978-3-662-00367-1 ISBN 978-3-7908-1832-1 (eBook)
DOI 10.1007/978-3-7908-1832-1

Physica-Verlag Heidelberg New York
a member of BertelsmannSpringer Science+Business Media GmbH

Softcover reprint of the hardcover 1st edition 2001

Hardcover Design: Erich Kirchner, Heidelberg

SPIN 10784664 88/2202-5 4 3 2 1 0 – Printed on acid-free paper

Preface

This book presents a unified view of modelling, simulation, and control of non-linear dynamical systems using soft computing techniques and fractal theory. Our particular point of view is that modelling, simulation, and control are problems that cannot be considered apart, because they are intrinsically related in real world applications. Control of non-linear dynamical systems cannot be achieved if we don't have the appropriate model for the system. On the other hand, we know that complex non-linear dynamical systems can exhibit a wide range of dynamic behaviors (ranging from simple periodic orbits to chaotic strange attractors), so the problem of simulation and behavior identification is a very important one. Also, we want to automate each of these tasks because in this way it is more easy to solve a particular problem. A real world problem may require that we use modelling, simulation, and control, to achieve the desired level of performance needed for the particular application.

Soft computing consists of several computing paradigms, including fuzzy logic, neural networks, evolutionary computation, and chaos theory, which can be used to produce powerful hybrid intelligent systems. We believe that to really be able to automate modelling, simulation, and control of dynamical systems, we require the use of hybrid combinations of soft computing techniques. In this way, we can exploit the advantages that each technique offers for solving these difficult problems. On the other hand, fractal theory provides us with powerful mathematical tools that can be used to understand the geometrical complexity of natural or computational objects. We believe that, in many cases, it is necessary to use fractal techniques to understand the geometry of the problem at hand.

This book is intended to be a major reference for scientists and engineers interested in applying new computational and mathematical tools to modelling, simulation, and control of non-linear dynamical systems. This book can also be used as a textbook or major reference for graduate courses like: soft computing, control of dynamical systems, applied artificial intelligence, and similar ones. We consider that this book can also be used to get novel ideas for new lines of research, or to continue the lines of research proposed by the authors of the book.

In Chapter one, we begin by giving a brief introduction to the problems of modelling, simulation, and control of non-linear dynamical systems. We discuss the importance of solving these problems for real-world applications. We motivate the reasons for automating modelling, simulation, and control using computational techniques. We also outline the importance of using soft computing techniques and fractal theory to really achieve automated modelling, simulation, and adaptive control of non-linear dynamical systems.

We describe in Chapter 2 the main ideas underlying fuzzy logic, and the application of this powerful computational theory to the problems of modelling and control of dynamical systems. We discuss in some detail fuzzy set theory, fuzzy reasoning, and fuzzy inference systems. We also describe briefly the generalization of conventional (type-1) fuzzy logic to what is now known as type-2 fuzzy logic. At the end, we also give some general guidelines for the process of fuzzy modelling and control. The importance of fuzzy logic as a basis for developing intelligent systems for control has been recognized in several areas of application. For this reason, we consider this chapter essential to understand the new methods for modelling, simulation, and control that are described in subsequent chapters.

We describe in Chapter 3 the basic concepts, notation, and the learning algorithms for neural networks. We discuss in some detail feedforward networks, adaptive neuro-fuzzy inference systems, neuro-fuzzy control, and adaptive neuro-control. First, we give a brief review of the basic concepts of neural networks and the backpropagation learning algorithm. We then continue with a general description of adaptive neuro-fuzzy systems. Finally, we end the chapter with a review of the most important current methods for neuro-control, and some general remarks about adaptive control and model-based control. The importance of neural networks as a computational tool to achieve "intelligence" for software systems has been well recognized in the literature of the area. For this reason, neural networks have been applied for solving complex problems of modelling, identification and control.

We describe in Chapter 4 the basic concepts and notation of genetic algorithms, and simulated annealing. We also describe the application of genetic algorithms for evolving neural networks, and fuzzy systems. Both genetic algorithms and simulated annealing are basic search methodologies that can be used for modelling and simulation of complex non-linear dynamical systems. Since both techniques can be considered as general purpose optimization methodologies, we can use any of them to find the model which minimizes the fitting error for a specific problem. Also, genetic algorithms can be used to automate the simulation of dynamical systems, because it can be used to produce the best set of parameter values for a model of the system. As genetic algorithms are based on the ideas of natural evolution, we can use this methodology to try evolving a neural network or fuzzy system for a particular application. The problem of finding the best architecture of a neural network is very important because there are no theoretical results on this, and in many cases we are forced to trial and error unless we use a genetic algorithm to automate this process. A similar thing occurs in the determining the optimum number of rules and membership functions of a fuzzy system for a particular application, here a genetic algorithm can also help us avoid time consuming trial and error.

We describe in Chapter 5 the basic concepts of dynamical systems and fractal theory, which are two powerful mathematical theories that enable the understanding of complex non-linear phenomena. We also describe the general mathematical methods for controlling chaos in dynamical systems. Dynamical systems theory gives us the general framework for studying non-linear systems. It can also be used to for behavior identification in complex non-linear dynamical systems. On the other hand, fractal theory gives us powerful concepts and techniques that can be used to measure the complexity of geometrical objects. In particular, the concept of the fractal dimension is very useful in classifying the complexity for time series of measured data for a problem. We discuss at the end of the chapter the problem of controlling chaotic behavior in non-linear dynamical systems. We review several methods for chaos control based on different ideas of how to move from a chaotic orbit of the dynamical system to a periodic stable orbit. This is very important in real world applications, because in many cases we need to control chaos to avoid physical damage to the system.

We describe in Chapter 6 our new method for time series analysis and prediction. This method is based on a new hybrid fuzzy-fractal approach, that combines the advantages of the fractal dimension for measuring the complexity of the time series, and of fuzzy logic for constructing a set of fuzzy rules to model the problem. We also define a new concept, which we have called the fuzzy fractal dimension, to generalize the mathematical definition of the capacity dimension. We show results of the application of our new method to real time series and measure the efficiency of our new hybrid approach for modelling. This new approach for time series prediction can be very useful for forecasting the behavior of complex non-linear dynamical systems.

We describe in Chapter 7 a new method for modelling complex dynamical systems using multiple differential equations. This method is a new fuzzy reasoning procedure that can be considered as a generalization of Sugeno's original fuzzy inference system. Our method uses a set of fuzzy rules, which have as consequents non-linear differential equations. Each equation is viewed as local model, for each region of the domain of definition for a complex non-linear dynamical system. The general idea of the approach is to simplify the task of modelling complex dynamical systems, by dividing the domain in smaller regions in which a simpler model can be formulated. We show the application of our new approach to the problems of modelling complex robotic dynamic systems, and aircraft systems. Modelling these dynamical systems can be used in controlling their complex non-linear dynamic behavior.

We describe in Chapter 8 a new method for automated simulation of non-linear dynamical systems with a hybrid fuzzy-genetic approach. Genetic algorithms are used, in this case, to generate the parameter values for the mathematical models of the dynamical systems. Fuzzy logic is used to model the uncertainty in behavior identification for a particular dynamical system. A set of fuzzy rules can be developed as a classification scheme of the dynamic behaviors using as information the fractal dimension or the Lyapunov exponents of the system. We also present a new concept, which we have called fuzzy chaos, to generalize the mathematical definition of chaos. In many cases, due to uncertainty it is more appropriate to find fuzzy regions of specific dynamic behaviors, even more for the complex chaotic behavior. We show results of the application of this hybrid fuzzy-genetic approach to the problem of automated simulation for robotic dynamic systems.

We present in Chapter 9 a new method for adaptive model-based control of robotic dynamic systems. This method combines the use of neural networks, fuzzy logic, and fractal theory to achieve real time control of robotic systems. Our neuro-fuzzy-fractal approach uses neural networks for identification and control, fuzzy logic for modelling, and fractal theory for time series analysis. We show results of our hybrid approach for several types of robot manipulators. Robotic systems are highly non-linear dynamical systems with a wide range of dynamic behaviors going from simple periodic behavior to the completely unstable chaotic behavior. In this case, chaotic behavior has to be avoided to prevent physical damage to the robot, and also other types of unstable behavior could be dangerous for the system. For this reason, the control of these systems is very important in real world applications. Our hybrid neuro-fuzzy-fractal approach exploits the advantages that each technique has for achieving the ultimate goal of controlling robotic dynamic systems in an efficient way.

We present in Chapter 10 the application of our new method for adaptive model-based control to the case of controlling biochemical reactors in the food industry. We use our hybrid neuro-fuzzy-fractal approach for controlling the complex behavior of biochemical reactors during production. Bioreactors are used in food production plants to produce the food with the required characteristics and level of quality. In this case, we need to control the reactor to optimize the production and the quality of the food product. Biochemical reactors use specific bacteria to produce chemical compounds that needed to obtain particular food products. The behavior of these reactors is highly non-linear and requires complex control strategies. For this reason, the application of soft computing techniques can help in achieving the goal of adaptive control of this type of reactors. In this case, neural networks are used for identification and control, fuzzy logic for modelling, and fractal theory for identifying bacteria during the production process.

We describe in Chapter 11 the application of soft computing techniques to the problem of controlling complex electrochemical processes. Electrochemical processes, like the ones used in battery formation, are highly non-linear and difficult to control. Also, mathematical models of these processes are difficult to obtain. The ultimate goal, in this case, is to control the process to optimize the manufactured product, avoiding at the same time going over the limiting temperature value for the electrochemical reaction. We use a hybrid neuro-fuzzy-genetic approach to control the electrochemical process during battery formation in a manufacturing plant. Neural networks are used for modelling the electrochemical reaction, fuzzy logic is used for controlling the process, and genetic algorithms are used to optimize the membership functions for the fuzzy systems using as input the measured data for the process.

We describe in Chapter 12 the application of soft computing techniques to the problem of controlling aircraft dynamic systems. Aircraft systems, are very complicated non-linear dynamical systems that show a wide range of dynamic behaviors even chaos. For this reason, controlling these systems is a very difficult task. We use a hybrid neuro-fuzzy-fractal approach for controlling the aircraft dynamics during flight. Neural networks are used for identification and control of the system, fuzzy logic for modelling, and fractal theory to measure the complexity of dynamic situation. We use our new fuzzy reasoning procedure for multiple differential equations to model the complex dynamical system. Our hybrid neuro-fuzzy-fractal approach enables on-line real time control of these type of dynamical systems.

Finally, we present in Chapter 13 the application of soft computing techniques to the problem of controlling dynamic economic systems. We consider the complex situation of the competing economies of three countries with international trade. This economic system is highly non-linear and coupled, and for this reason has a wide range of dynamic behaviors going from simple stable periodic orbits to the very unstable chaotic behavior. The ultimate goal, in this case, is to control international trade so as achieve stable economic growth and optimize the national income. We can use mathematical models of this economic system to simulate different kinds of behaviors and analyze the possible routes for control. We can then use a fuzzy system with rules having as consequents differential equations, to completely model the economic dynamic system. We can also use neural networks to control the economic system. The neural networks can be trained with historical data or using a genetic algorithm and simulations. Our hybrid approach can enable the control of this complex economic dynamic system, and illustrates that computing techniques can also be applied to problems in economics.

Contents

Chapter 1

Introduction to Control of Non-Linear Dynamical Systems

We describe in this book, new methods for modelling, simulation, and control of dynamical systems using soft computing techniques and fractal theory. Soft Computing (SC) consists of several computing paradigms, including fuzzy logic, neural networks, and genetic algorithms, which can be used to produce powerful hybrid intelligent systems. Fractal theory provides us with the mathematical tools to understand the geometrical complexity of natural objects and can be used for identification and modelling purposes. Combining SC techniques with fractal theory, we can take advantage of the "intelligence" provided by the computer methods and also take advantage of the descriptive power of fractal mathematical tools. Non-linear dynamical systems can exhibit extremely complex dynamic behavior, and for this reason, it is of great importance to develop intelligent computational tools that will enable the identification of the best model for a particular dynamical system, then obtaining the best simulations for the system, and also achieving the goal of controlling the dynamical system in a desired manner.

As a prelude, we provide a brief overview of the existing methodologies for modelling, simulation, and control of dynamical systems. We then show our own approach in dealing with these problems. Our particular point of view is that modelling, simulation and control are problems that can not be considered apart because they are intrinsically related in real-world applications. We think that in many cases control of non-linear dynamical systems can not be achieved if we don't have proper mathematical models for the systems. Also, useful simulations of a model, that can give us numerical insights into the behavior of a dynamical system, can not be obtained if we don't have the appropriate model.

Traditionally, mathematical models of dynamical systems have been obtained by statistical methods, which lack the accuracy needed in real-world applications. We instead of the traditional approach, consider a general modelling method using fuzzy logic techniques. Our modelling approach consists in a set of fuzzy if-then rules that relate a specific mathematical model to a specific region of the universe of discourse. We have proposed a new reasoning methodology for multiple differential equations that can be considered as a generalization of Sugeno's original approach. In Sugeno's original approach the fuzzy if-then rules have polynomials in the consequent part. Instead of simple polynomials, we are now using non-linear differential equations. The idea of our approach is to use relatively simple models for each region of the universe of discourse, instead of trying to use only one complicated model for the complete system. We have applied this approach to the case of robotic systems, aircraft systems, and biochemical reactors with excellent results. In all of these cases, the dynamical systems have very complicated behaviors and for this reason it is very difficult to develop a unique model for the system. Instead, we use local models for each region of the universe of discourse and relate them by fuzzy rules to the necessary conditions for the application of the models.

The problem of numerical simulation of non-linear dynamical systems consists in the application of an appropriate mapping for different parameter values, and behavior identification according to the time series generated in the simulation. Traditionally, the simulation of a specific dynamical system is done by, manually trying different sets of parameter values for the mathematical model, and then checking the dynamic behaviors corresponding to these parameter values. The problem of this approach is that is time consuming and in many cases interesting behaviors are never explored. We have used SC techniques to automate the process of parameter value selection for a mathematical model and then the subsequent identification of the limiting behavior of the system. We use genetic algorithms to evolve a population of parameter values and find at the end the best ones for identification. Once the mapping is iterated with the best parameter values, a set of fuzzy rules is used for behavior identification. The set of fuzzy if-then rules represents the expert knowledge for dynamic behavior identification. We also use the concept of the fractal dimension as a measure of the complexity of the time series generated by the iteration of the dynamical system. The fuzzy-fractal-genetic approach for automated simulation has been applied to different types of dynamical systems with excellent results. As an example, we can mention that for robotic dynamic systems we can automatically find out interesting parameter values in the sense that interesting behavior is explored. These simulations are useful in finding out how to control a complex dynamical system.

The problem of controlling non-linear dynamical systems consists in finding out control laws to achieve a desired trajectory for the system. Of course, control laws can be deduced from the mathematical models of the dynamical system or from computational models that use real data for the problem. Traditionally, the control of dynamical systems has been considered as a linear control problem, i.e., linear models are used to represent the dynamical system and accordingly linear control laws are obtained. Of course, many real world problems have been solved to a certain degree with this approach. However, real world dynamical systems are intrinsically non-linear in nature, and assuming that they are linear is only a very crude approximation of reality. We have to say here that some researchers are trying right now to generalize the results of classical linear control theory to the case of non-linear dynamical systems. However, results have been difficult to achieve to the moment. Another approach is to use computational models to represent the non-linear dynamical system and the control laws needed to achieve the desired level of performance in real world applications. Our approach is to use SC techniques and fractal theory to control non-linear dynamical systems in real time. We use fuzzy logic for modelling the dynamical system with a set of fuzzy rules relating the models (as differential equations) to their corresponding regions in the universe of discourse. On the other hand, we use neural networks for control of the dynamical system using as a reference model the one given by the fuzzy system for modelling. Of course, the neural network has to be trained with real control data for the specific problem. We can also use neural networks for identification of the parameters of the specific models used in the fuzzy rule base mentioned above. In many cases, we also use the fractal dimension as a measure of classification for the complexity of the problem at hand. For example, for the case of biochemical reactors we use the fractal dimension to classify bacteria by the geometrical complexity of their colonies. In this way, the fractal dimension is used to control the process of production because bacteria are responsible for the quality of the food being produced. The neuro-fuzzy-fractal hybrid approach for controlling non-linear dynamical systems uses the best that each technique has to offer for these types of applications. Of course, other SC techniques can also be used in controlling non-linear dynamical systems, for example in the hybrid approach mentioned above, we can also use genetic algorithms to find the best architecture for the neural networks, i.e., the best number of nodes and layers of the neural network. This genetic approach can help in the design of the intelligent control system for a particular application.

We illustrate these ideas in this book with several real world applications. We consider the problem of controlling robotic dynamic systems with our neuro-fuzzy-fractal approach. The problem of robot control is how to make the system follow a pre-specified desired trajectory while satisfying certain constraints. In this case, fuzzy logic is used for modelling these dynamical systems, and neural networks are used for control and identification. We also consider the use of a fuzzy-genetic approach for automated simulation of robotic systems. On the other hand, we consider the application of our neuro-fuzzy-fractal approach for controlling aircraft dynamic systems. This is done similarly to the case of robots, the main difference is in the type of models that we need to use for aircraft dynamic systems. We can also use the fuzzy-genetic approach for automated simulation of aircraft systems if we want to explore the diversity of dynamic behaviors possible for this type of systems. Also, we consider the application of our hybrid neuro-fuzzy-fractal approach for the case of controlling biochemical reactors used for food production. In this case, mathematical models are used to represent the population of bacteria in the bioreactor and the chemical compound been produced by the bacteria. The problem here is how to control conditions in the bioreactor in such a way as to optimize food production. We also consider the problem of controlling electrochemical processes as the ones used in battery formation. In this case, we have used a hybrid neuro-fuzzy-genetic approach for intelligent control of the electrochemical process. The neural network is used for modelling the process, fuzzy logic is used for control, and genetic algorithms for evolving the fuzzy system. Of course, this is another hybrid architecture for control of non-linear dynamical systems, but in this case we have found that this is the most appropriate one. Finally, we also consider the problem of controlling international trade between countries. This application comes from the field of economics and is very important for planning the economy of a specific country. The problem here is how to control the trade between countries in such a way as to optimize national income and achieve stable growth of the country. We have shown that international trade can produce chaotic unstable behavior of a system of at least three countries under certain specific conditions. Of course, we want to avoid this type of erratic dynamic behavior in the economy of the countries involved. We have used a hybrid approach combining SC techniques for controlling dynamic behavior in this type of economic systems.

The diversity of the applications considered in this book, gives idea of the universality of the hybrid approach of combining SC techniques for controlling non-linear dynamical systems. The best combination of SC techniques may change because of the properties of the system under consideration, but one can always find the hybrid architecture needed for achieving the ultimate goal of control. Of course, we still need to do a lot of work in finding out general rules for knowing in advance the best combination of techniques. The best architecture of the intelligent control system has to be determined in many cases by a lot experimental work. But at the end the rewards of this experimental work are satisfying.

Chapter 2

Fuzzy Logic

This chapter introduces the basic concepts, notation, and basic operations for fuzzy sets that will be needed in the following chapters. Since research on Fuzzy Set Theory has been underway for over 30 years now, it is practically impossible to cover all aspects of current developments in this area. Therefore, the main goal of this chapter is to provide an introduction to and a summary of the basic concepts and operations that are relevant to the study of fuzzy sets. We also introduce in this chapter the definition of linguistic variables and linguistic values and explain how to use them in fuzzy rules, which are an efficient tool for quantitative modelling of words or sentences in a natural or artificial language. By interpreting fuzzy rules as fuzzy relations, we describe different schemes of fuzzy reasoning, where inference procedures based on the concept of the compositional rule of inference are used to derive conclusions from a set of fuzzy rules and known facts. Fuzzy rules and fuzzy reasoning are the basic components of fuzzy inference systems, which are the most important modelling tool, based on fuzzy set theory.

The "fuzzy inference system" is a popular computing framework based on the concepts of fuzzy set theory, fuzzy if-then rules, and fuzzy reasoning (Jang, Sun & Mizutani, 1997). It has found successful applications in a wide variety of fields, such as automatic control, data classification, decision analysis, expert systems, time series prediction, robotics, and pattern recognition (Jamshidi, 1997). Because of its multidisciplinary nature, the fuzzy inference system is known by numerous other names, such as "fuzzy expert system" (Kandel, 1992), "fuzzy model" (Sugeno & Kang, 1988), "fuzzy associative memory" (Kosko, 1992), and simply "fuzzy system".

The basic structure of a fuzzy inference system consists of three conceptual components: a "rule base", which contains a selection of fuzzy rules; a "data base" (or "dictionary"), which defines the membership functions used in the fuzzy rules; and a "reasoning mechanism", which performs the inference procedure upon the rules and given facts to derive a reasonable output or conclusion. In general, we can say that a fuzzy inference system implements a non-linear mapping from its input space to output space. This mapping is accomplished by a number of fuzzy if-then rules, each of which describes the local behavior of the mapping. In particular, the antecedent of a rule defines a fuzzy region in the input space, while the consequent specifies the output in the fuzzy region.

We also describe very briefly a new area in fuzzy logic, which studies type-2 fuzzy sets and type-2 fuzzy systems. Basically, a type-2 fuzzy set is a set in which we also have uncertainty about the membership function. Since we are dealing with uncertainty for the conventional fuzzy sets (which are called type-1 fuzzy sets here) we can achieve a higher degree of approximation in modelling real world problems. Of course, type-2 fuzzy systems consist of fuzzy if-then rules, which contain type-2 fuzzy sets. We can say that type-2 fuzzy logic is a generalization of conventional fuzzy logic (type-1) in the sense that uncertainty is not only limited to the linguistic variables but also is present in the definition of the membership functions.

In what follows, we shall first introduce the basic concepts of fuzzy sets, and fuzzy reasoning. Then we will introduce and compare the three types of fuzzy inference systems that have been employed in various applications. We will also consider briefly type-2 fuzzy logic systems and the comparison to type-1 fuzzy systems. Finally, we will address briefly the features and problems of fuzzy modelling, which is concerned with the construction of fuzzy inference systems for modelling a given target system.

2.1 Fuzzy Set Theory

Let X be a space of objects and x be a generic element of X. A classical set A, $A \subseteq X$, is defined by a collection of elements or objects $x \in X$, such that each x can either belong or not belong to the set A. By defining a "characteristic function" for each element $x \in X$, we can represent a classical set A by a set of order pairs (x,0) or (x,1), which indicates $x \notin A$ or $x \in A$, respectively.

Unlike the aforementioned conventional set, a fuzzy set (Zadeh, 1965) expresses the degree to which an element belong to a set. Hence the characteristic function of a fuzzy set is allowed to have values between 0 and 1, which denotes the degree of membership of an element in a given set.

<u>Definition</u> 2.1 **Fuzzy sets and membership functions**
If X is a collection of objects denoted generically by x, then a "fuzzy set" A in X is defined as a set of ordered pairs:

$$A = \{(x, \mu_A(x)) \mid x \in X\}. \tag{2.1}$$

where $\mu_A(x)$ is called "membership function" (or MF for short) for the fuzzy set A. The MF maps each element of X to a membership grade (or membership value) between 0 and 1.

Obviously, the definition of a fuzzy set is a simple extension of the definition of a classical set in which the characteristic function is permitted to have any values between 0 and 1. If the values of the membership function $\mu_A(x)$ is restricted to either 0 or 1, then A is reduced to a classical set and $\mu_A(x)$ is the characteristic function of A.

A fuzzy set is uniquely specified by its membership function. To describe membership functions more specifically, we shall define the nomenclature used in the literature (Jang, Sun & Mizutani, 1997).

Definition 2.2 **Support**
The "support" of a fuzzy set A is the set of all points x in X such that $\mu_A(x) > 0$:

$$\text{support}(A) = \{ x \mid \mu_A(x) > 0 \}. \tag{2.2}$$

Definition 2.3 **Core**
The "core" of a fuzzy set is the set of all points x in X such that $\mu_A(x) = 1$:

$$\text{core}(A) = \{ x \mid \mu_A(x) = 1 \}. \tag{2.3}$$

Definition 2.4 **Normality**
A fuzzy set A is "normal" if its core is nonempty. In other words, we can always find a point $x \in X$ such that $\mu_A(x) = 1$.

Definition 2.5 **Crossover points**
A "crossover point" of a fuzzy set A is a point $x \in X$ at which $\mu_A(x) = 0.5$:

$$\text{crossover}(A) = \{ x \mid \mu_A(x) = 0.5 \}. \tag{2.4}$$

Definition 2.6 **Fuzzy singleton**
A fuzzy set whose support is a single point in X with $\mu_A(x) = 1$ is called a "fuzzy singleton".

Corresponding to the ordinary set operations of union, intersection and complement, fuzzy sets have similar operations, which were initially defined in Zadeh's seminal paper (Zadeh, 1965). Before introducing these three fuzzy set operations, first we shall define the notion of containment, which plays a central role in both ordinary and fuzzy sets. This definition of containment is, of course, a natural extension of the case for ordinary sets.

Definition 2.7 **Containment**

The fuzzy set A is "contained" in fuzzy set B (or, equivalently, A is a "subset" of B) if and only if $\mu_A(x) \leq \mu_B(x)$ for all x. Mathematically,

$$A \subseteq B \Leftrightarrow \mu_A(x) \leq \mu_B(x). \quad (2.5)$$

Definition 2.8 **Union**

The "union" of two fuzzy sets A and B is a fuzzy set C, written as $C = A \cup B$ or $C = A$ OR B, whose MF is related to those of A and B by

$$\mu_C(x) = \max(\ \mu_A(x),\ \mu_B(x)\ ) = \mu_A(x) \vee \mu_B(x). \quad (2.6)$$

Definition 2.9 **Intersection**

The "intersection" of two fuzzy sets A and B is a fuzzy set C, written as $C = A \cap B$ or $C = A$ AND B, whose MF is related to those of A and B by

$$\mu_C(x) = \min(\ \mu_A(x),\ \mu_B(x)\ ) = \mu_A(x) \wedge \mu_B(x). \quad (2.7)$$

Definition 2.10 **Complement or Negation**

The "complement" of a fuzzy set A, denoted by $\bar{A}$ ($\neg$A, NOT A), is defined as

$$\mu_{\bar{A}}(x) = 1 - \mu_A(x). \quad (2.8)$$

As mentioned earlier, a fuzzy set is completely characterized by its MF. Since most fuzzy sets in use have a universe of discourse X consisting of the real line R, it would be impractical to list all the pairs defining a membership function. A more convenient and concise way to define a MF is to express it as a mathematical formula. First we define several classes of parameterized MFs of one dimension.

Definition 2.11 **Triangular MFs**

A "triangular MF" is specified by three parameters {a, b, c} as follows:

$$y = \text{triangle}(x;a,b,c) = \begin{cases} 0, & x \leq a. \\ (x-a)/(b-a), & a \leq x \leq b. \\ (c-x)/(c-b), & b \leq x \leq c. \\ 0, & c \leq x. \end{cases} \quad (2.9)$$

The parameters {a,b,c} (with a < b < c) determine the x coordinates of the three corners of the underlying triangular MF.

Figure 2.1 (a) illustrates a triangular MF defined by triangle(x; 10, 20, 40).

Definition 2.12 **Trapezoidal MFs**
A "trapezoidal MF" is specified by four parameters {a, b, c, d} as follows:

$$\text{trapezoid}\,(x;a,b,c,d) = \begin{cases} 0, & x \le a\,. \\ (x-a)/(b-a), & a \le x \le b\,. \\ 1, & b \le x \le c\,. \\ (d-x)/(d-c), & c \le x \le d\,. \\ 0, & d \le x\,. \end{cases} \qquad (2.10)$$

The parameters {a, b, c, d} (with $a < b \le c < d$) determine the x coordinates of the four corners of the underlying trapezoidal MF. Figure 2.1 (b) illustrates a trapezoidal MF defined by trapezoid(x; 10, 20 40, 75).

Due to their simple formulas and computational efficiency, both triangular MFs and trapezoidal MFs have been used extensively, especially in real-time implementations. However, since the MFs are composed of straight line segments, they are not smooth at the corner points specified by the parameters. In the following we introduce other types of MFs defined by smooth and nonlinear functions.

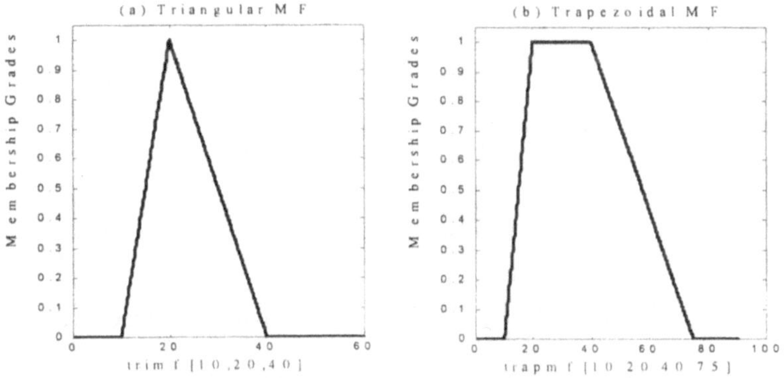

(a) Triangular MF b) Trapezoidal MF

Figure 2.1 Examples of two types of parameterized MFs

Definition 2.13 **Gaussian MFs**
A "Gaussian MF" is specified by two parameters {c , σ }:

$$\text{gaussian}(x; c, \sigma) = e^{-\frac{1}{2}\left(\frac{x-c}{\sigma}\right)^2} \qquad (2.11)$$

A "Gaussian MF is determined completely by c and σ ; c represents the MFs center and σ determines the MFs width. Figure 2.2 (a) plots a Gaussian MF defined by gaussian (x; 50, 20).

<u>Definition</u> 2.14 **Generalized bell MFs**
A "generalized bell MF" is specified by three parameters {a, b, c}:

$$\text{bell}(x; a, b, c) = \frac{1}{1 + |(x-c)/a|^{2b}} \qquad (2.12)$$

where the parameter b is usually positive. We can note that this MF is a direct generalization of the Cauchy distribution used in probability theory, so it is also referred to as the "Cauchy MF". Figure 2.2 (b) illustrates a generalized bell MF defined by bell(x; 20, 4, 50).

Although the Gaussian MFs and bell MFs achieve smoothness, they are unable to specify asymmetric MFs, which are important in certain applications. Next we define the sigmoidal MF, which is either open left or right.

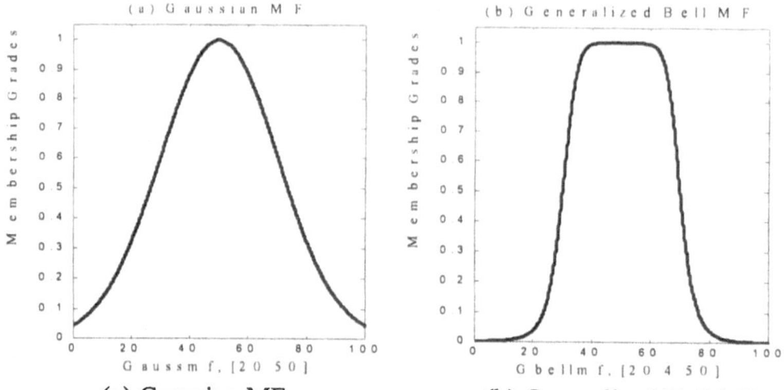

(a) Gaussian MF (b) Generalized Bell MF

Figure 2.2 Examples of two classes of parameterized continuous MFs.

<u>Definition</u> 2.15 **Sigmoidal MFs**
A "Sigmoidal MF" is defined by the following equation:

$$\text{sig}(x; a, c) = \frac{1}{1 + \exp[-a(x-c)]} \qquad (2.13)$$

where a controls the slope at the crossover point x = c.

Depending on the sign of the parameter "a", a sigmoidal MF is inherently open right or left and thus is appropriate for representing concepts such as "very large" or "very negative". Figure 2.3 shows two sigmoidal functions y_1 =sig(x; 1, -5) and y_2 =sig(x; -2, 5).

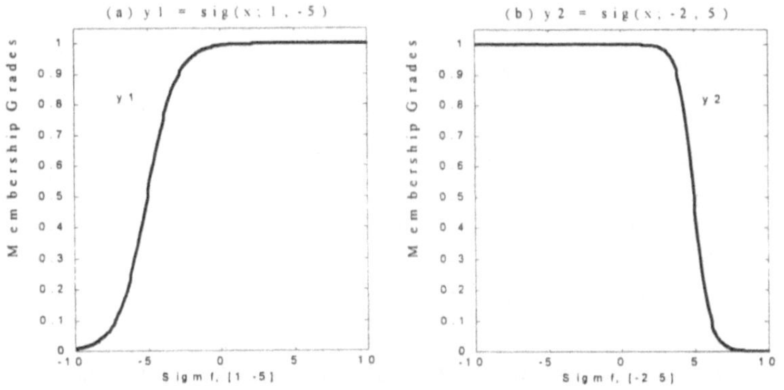

(a) y_1 = sig(x; 1, -5) (b) y_2 = sig(x; -2, 5)

Figure 2.3 Two sigmoidal functions y_1 and y_2 .

2.2 Fuzzy Reasoning

As was pointed out by Zadeh in his work on this area (Zadeh, 1973), conventional techniques for system analysis are intrinsically unsuited for dealing with humanistic systems, whose behavior is strongly influenced by human judgment, perception, and emotions. This is a manifestation of what might be called the "principle of incompatibility": "As the complexity of a system increases, our ability to make precise and yet significant statements about its behavior diminishes until a threshold is reached beyond which precision and significance become almost mutually exclusive characteristics" (Zadeh, 1973). It was because of this belief that Zadeh proposed the concept of linguistic variables (Zadeh, 1971) as an alternative approach to modelling human thinking.

Definition 2.16 **Linguistic variables**
A "Linguistic variable" is characterized by a quintuple (x, T(x), X, G, M) in which x is the name of the variable; T(x) is the "term set" of x-that is, the set of its "linguistic values" or "linguistic terms"; X is the universe of discourse, G is a "syntactic rule" which generates the terms in T(x); and M is a "semantic rule" which associates with each linguistic value A its meaning M(A), where M(A) denotes a fuzzy set in X.

Definition 2.17 **Concentration and dilation of linguistic values**
Let A be a linguistic value characterized by a fuzzy set membership function $\mu_A(.)$. Then A^k is interpreted as a modified version of the original linguistic value expressed as

$$A^k = \int_X [\,\mu_A(x)]^k / x \ . \tag{2.14}$$

In particular, the operation of "concentration" is defined as

$$\text{CON}\,(A) = A^2 \ , \tag{2.15}$$

while that of "dilatation" is expressed by

$$\text{DIL}\,(A) = A^{0.5} \ . \tag{2.16}$$

Conventionally, we take CON(A) and DIL(A) to be the results of applying the hedges "very" and "more or less", respectively, to the linguistic term A. However, other consistent definitions for these linguistic hedges are possible and well justified for various applications.

Following the definitions given before, we can interpret the negation operator NOT and the connectives AND and OR as

$$\begin{aligned} \text{NOT}(A) &= \neg A = \int_X [\,1 - \mu_A(x)\,] / x \ , \\ A \text{ AND } B &= A \cap B = \int_X [\,\mu_A(x) \wedge \mu_B(x)\,] / x \ , \\ A \text{ OR } B &= A \cup B = \int_X [\,\mu_A(x) \vee \mu_B(x)\,] / x \ . \end{aligned} \tag{2.17}$$

respectively, where A and B are two linguistic values whose meanings are defined by $\mu_A(.)$ and $\mu_B(.)$.

Definition 2.18 **Fuzzy If-Then Rules**
A "fuzzy if-then rule" (also known as "fuzzy rule", "fuzzy implication", or "fuzzy conditional statement") assumes the form

$$\textbf{if}\ x \text{ is } A\ \textbf{then}\ y \text{ is } B \ , \tag{2.18}$$

where A and B are linguistic values defined by fuzzy sets on universes of discourse X and Y, respectively. Often "x is A" is called "antecedent" or "premise", while "y is B" is called the "consequence" or "conclusion".

Examples of fuzzy if-then rules are widespread in our daily linguistic expressions, such as the following:

- If pressure is high, then volume is small.
- If the road is slippery, then driving is dangerous.
- If the speed is high, then apply the brake a little.

Before we can employ fuzzy if-then rules to model and analyze a system, first we have to formalize what is meant by the expression "if x is A then y is B", which is sometimes abbreviated as $A \rightarrow B$. In essence, the expression describes a relation between two variables x and y; this suggests that a fuzzy if-then rule is defined as a binary fuzzy relation R on the product space $X \times Y$. Generally speaking, there are two ways to interpret the fuzzy rule $A \rightarrow B$. If we interpret $A \rightarrow B$ as A "coupled with" B then

$$R = A \rightarrow B = A \times B = \int_{X \times Y} \mu_A(x) * \mu_B(y) / (x,y)$$

where * is an operator for intersection (Mamdani & Assilian, 1975). On the other hand, if $A \rightarrow B$ is interpreted as A "entails" B, then it can be written as one of two different formulas:

- Material implication:
$$R = A \rightarrow B = \neg A \cup B \ . \qquad (2.19)$$
- Propositional Calculus:
$$R = A \rightarrow B = \neg A \cup (A \cap B) \ . \qquad (2.20)$$

Although these two formulas are different in appearance, they both reduce to the familiar identity $A \rightarrow B \equiv \neg A \cup B$ when A and B are propositions in the sense of two-valued logic.

Fuzzy reasoning, also known as approximate reasoning, is an inference procedure that derives conclusions from a set of fuzzy if-then rules and known facts. The basic rule of inference in traditional two-valued logic is "modus ponens", according to which we can infer the truth of a proposition B from the truth of A and the implication $A \rightarrow B$. This concept is illustrated as follows:

premise 1 (fact):	x is A ,
premise 2 (rule):	if x is A then y is B ,
consequence (conclusion):	y is B .

However, in much of human reasoning, modus ponens is employed in an approximate manner. This is written as

premise 1 (fact):	x is A'
premise 2 (rule):	if x is A then y is B ,
consequence (conclusion):	y is B'

where A' is close to A and B' is close to B. When A, B, A' and B' are fuzzy sets of appropriate universes, the foregoing inference procedure is called "approximate reasoning" or "fuzzy reasoning"; it is also called "generalized modus ponens" (GMP for short), since it has modus ponens as a special case.

Definition 2.19 **Fuzzy reasoning**
Let A, A', and B be fuzzy sets of X, X, and Y respectively. Assume that the fuzzy implication A → B is expressed as a fuzzy relation R on X × Y. Then the fuzzy set B induced by "x is A'" and the fuzzy rule "if x is A then y is B" is defined by

$$\mu_{B}'(y) = \max_X \min [\ \mu_{A}'(x), \mu_R(x, y)\ ]$$
$$= \vee_X [\ \mu_{A}'(x) \wedge \mu_R(x, y)\ ]\ . \qquad (2.21)$$

Now we can use the inference procedure of fuzzy reasoning to derive conclusions provided that the fuzzy implication A → B is defined as an appropriate binary fuzzy relation.

Single Rule with Single Antecedent
This is the simplest case, and the formula is available in Equation (2.21). A further simplification of the equation yields

$$\mu_{B}'(y) = [\ \vee_X (\ \mu_{A}'(x) \wedge \mu_A(x)\ )\ ] \wedge \mu_B(y)$$
$$= \omega \wedge \mu_B(y)$$

In other words, first we find the degree of match ω as the maximum of $\mu_{A}'(x) \wedge \mu_A(x)$; then the MF of the resulting B' is equal to the MF of B clipped by ω. Intuitively, ω represents a measure of degree of belief for the antecedent part of a rule; this measure gets propagated by the if-then rules and the resulting degree of belief or MF for the consequent part should be no greater than ω.

Multiple Rules with Multiple Antecedents
The process of fuzzy reasoning or approximate reasoning for the general case can be divided into four steps:

1) Degrees of compatibility: Compare the known facts with the antecedents of fuzzy rules to find the degrees of compatibility with respect to each antecedent MF.
2) Firing Strength: Combine degrees of compatibility with respect to antecedent MFs in a rule using fuzzy AND or OR operators to form a firing strength that indicates the degree to which the antecedent part of the rule is satisfied.

3) Qualified (induced) consequent MFs: Apply the firing strength to the consequent MF of a rule to generate a qualified consequent MF.
4) Overall output MF: Aggregate all the qualified consequent MFs to obtain an overall output MF.

2.3 Fuzzy Inference Systems

The "Mamdani fuzzy inference system" (Mamdani & Assilian, 1975) was proposed as the first attempt to control a steam engine and boiler combination by a set of linguistic control rules obtained from experienced human operators. Figure 2.4 is an illustration of how a two-rule Mamdani fuzzy inference system derives the overall output z when subjected to two numeric inputs x and y.

In Mamdani's application, two fuzzy inference systems were used as two controllers to generate the heat input to the boiler and throttle opening of the engine cylinder, respectively, to regulate the steam pressure in the boiler and the speed of the engine. Since the engine and boiler take only numeric values as inputs, a defuzzifier was used to convert a fuzzy set to a numeric value.

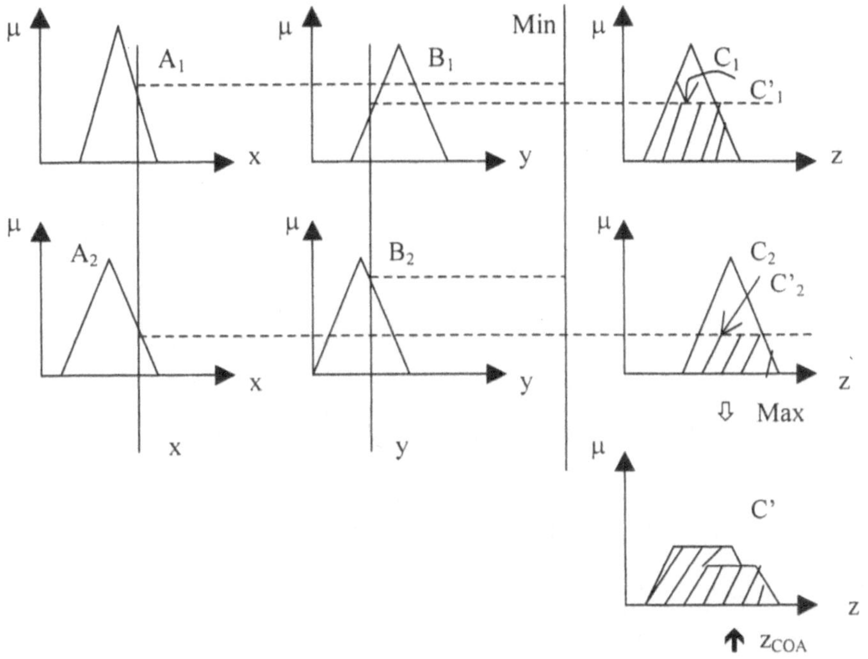

Figure 2.4 The Mamdani fuzzy inference system using the min and max operators.

Defuzzification

Defuzzification refers to the way a numeric value is extracted from a fuzzy set as a representative value. In general, there are five methods for defuzzifying a fuzzy set A of a universe of discourse Z, as shown in Figure 2.5 (Here the fuzzy set A is usually represented by an aggregated output MF, such as C' in Figure 2.4). A brief explanation of each defuzzification strategy follows.

- Centroid of area z_{COA}:

$$z_{COA} = \frac{\int_Z \mu_A(z) z dz}{\int_Z \mu_A(z) dz} \tag{2.22}$$

where $\mu_A(z)$ is the aggregated output MF. This is the most widely adopted defuzzification strategy, which is reminiscent of the calculation of expected values of probability distributions.

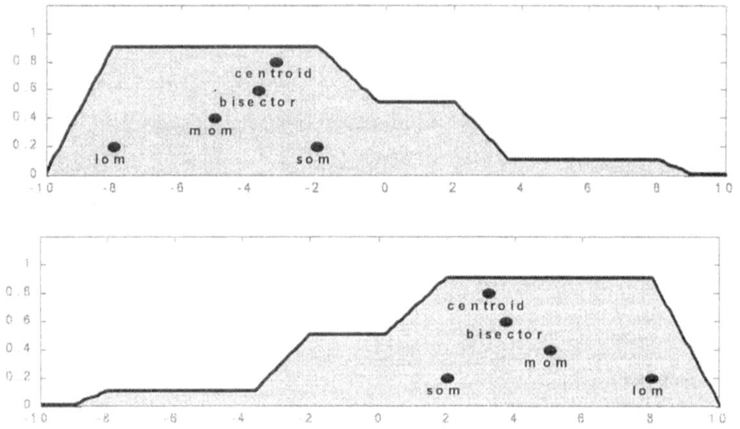

Figure 2.5 Various defuzzification methods for obtaining a numeric output.

- Bisector of area z_{BOA} : z_{BOA} satisfies

$$\int_\alpha^{z_{BOA}} \mu_A(z) dz = \int_{z_{BOA}}^{\beta} \mu_A(z) dz \ , \tag{2.23}$$

where $\alpha = \min\{z \mid z \in Z\}$ and $\beta = \max\{z \mid z \in Z\}$.

- Mean of maximum z_{MOM} : z_{MOM} is the average of the maximizing z at which the MF reach a maximum μ^*. Mathematically,

$$z_{MOM} = \frac{\int_{z'} z dz}{\int_{z'} dz}, \tag{2.24}$$

where $z' = \{ z \mid \mu_A(z) = \mu^* \}$. In particular, if $\mu_A(z)$ has a single maximum at $z = z^*$, then $z_{MOM} = z^*$. Moreover, if $\mu_A(z)$ reaches its maximum whenever $z \in [z_{left}, z_{right}]$ then $z_{MOM} = (z_{left} + z_{right}) / 2$.

- Smallest of maximum z_{SOM} : z_{SOM} is the minimum (in terms of magnitude) of the maximizing z.
- Largest of maximum z_{LOM} : z_{LOM} is the maximum (in terms of magnitude) of the maximizing z. Because of their obvious bias, z_{SOM} and z_{LOM} are not used as often as the other three defuzzification methods.

The calculation needed to carry out any of these five defuzzification operations is time-consuming unless special hardware support is available. Furthermore, these defuzzification operations are not easily subject to rigorous mathematical analysis, so most of the studies are based on experimental results. This leads to the propositions of other types of fuzzy inference systems that do not need defuzzification at all; two of them will be described in the following. Other more flexible defuzzification methods can be found in several more recent papers (Yager & Filer, 1993), (Runkler & Glesner, 1994).

Sugeno Fuzzy Models

The "Sugeno fuzzy model" (also known as the "TSK fuzzy model") was proposed by Takagi, Sugeno and Kang in an effort to develop a systematic approach to generating fuzzy rules from a given input-output data set (Takagi & Sugeno, 1985), (Sugeno & Kang, 1988). A typical fuzzy rule in a Sugeno fuzzy model has the form:

if x is A **and** y is B **then** $z = f(x,y)$

where A and B are fuzzy sets in the antecedent, while $z = f(x,y)$ is a traditional function in the consequent. Usually $f(x,y)$ is a polynomial in the input variables x and y, but it can be any function as long as it can appropriately describe the output of the model within the fuzzy region specified by the antecedent of the rule. When $f(x,y)$ is a first-order polynomial, the resulting fuzzy inference system is called a "first-order Sugeno fuzzy model". When f is constant, we then have a "zero-order Sugeno fuzzy model", which can be viewed either as a special case of the Mamdani inference system, in which each rule's consequent is specified by a fuzzy singleton, or a special case of the Tsukamoto fuzzy model (to be introduced next), in which each rule's consequent is specified by a MF of a step function center at the constant.

Figure 2.6 shows the fuzzy reasoning procedure for a first-order Sugeno model. Since each rule has a numeric output, the overall output is obtained via "weighted average", thus avoiding the time-consuming process of defuzzification required in a Mamdani model. In practice, the weighted average operator is sometimes replaced with the "weighted sum" operator (that is, $w_1z_1 + w_2z_2$ in Figure 2.6) to reduce computation further specially, in the training of a fuzzy inference system. However, this simplification could lead to the loss of MF linguistic meanings unless the sum of firing strengths (that is, Σw_i) is close to unity.

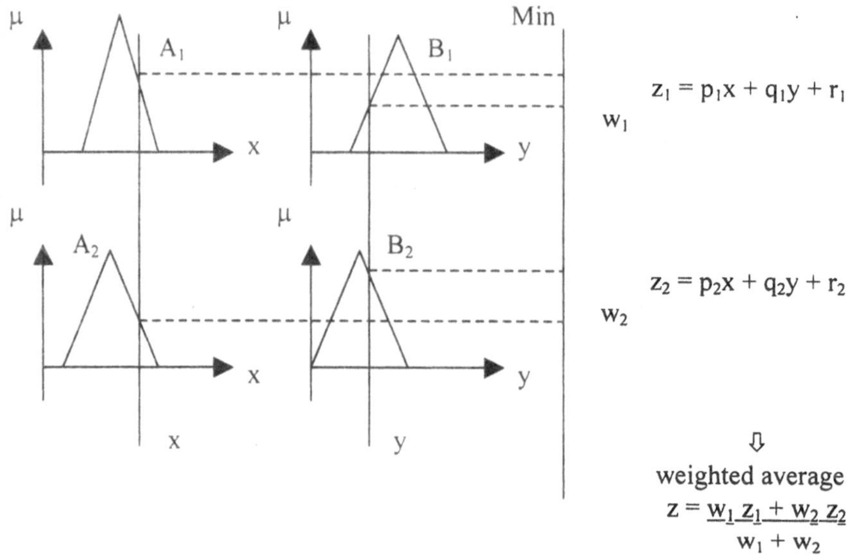

Figure 2.6 The Sugeno fuzzy model.

Tsukamoto Fuzzy Models

In the "Tsukamoto fuzzy models" (Tsukamoto, 1979), the consequent of each fuzzy if-then rule is represented by a fuzzy set with a monotonical MF, as shown in Figure 2.7. As a result, the inferred output of each rule is defined as a numeric value induced by the rule firing strength. The overall output is taken as the weighted average of each rule's output. Figure 2.7 illustrates the reasoning procedure for a two-input two-rule system.

Since each rule infers a numeric output, the Tsukamoto fuzzy model aggregates each rule's output by the method of weighted average and thus avoids the time-consuming process of defuzzification. However, the Tsukamoto fuzzy model is not used often since it is not as transparent as either the Mamdani or Sugeno fuzzy models.

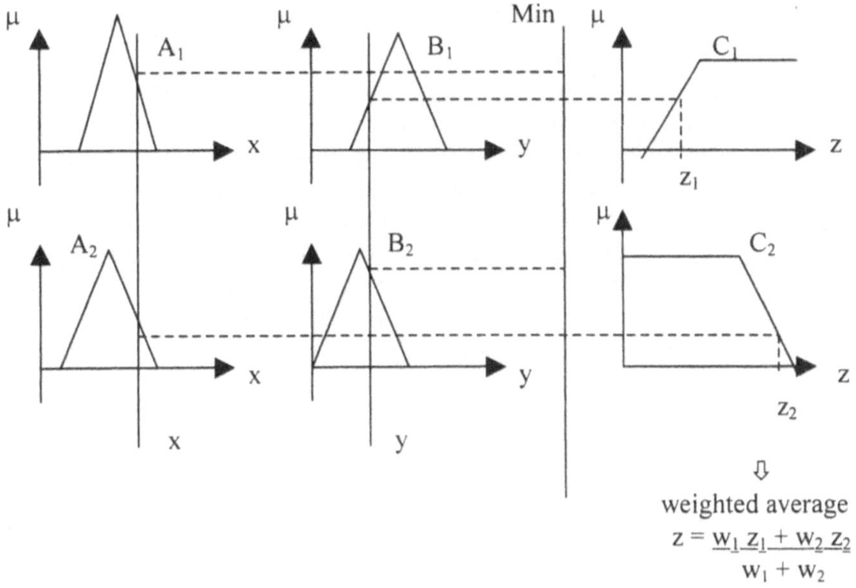

Figure 2.7 The Tsukamoto fuzzy model.

There are certain common issues concerning all the three fuzzy inference systems introduced previously, such as how to partition an input space and how to construct a fuzzy inference system for a particular application.

<u>Input Space Partitioning</u>

Now it should be clear that the spirit of fuzzy inference systems resembles that of "divide and conquer" - the antecedent of a fuzzy rule defines a local fuzzy region, while the consequent describes the behavior within the region via various constituents. The consequent constituent can be a consequent MF (Mamdani and Tsukamoto fuzzy models), a constant value (zero-order Sugeno model), or a linear equation (first-order Sugeno model). Different consequent constituents result in different fuzzy inference systems, but their antecedents are always the same. Therefore, the following discussion of methods of partitioning input spaces to form the antecedents of fuzzy rules is applicable to all three types of fuzzy inference systems.

- Grid partition: This partition method is often chosen in designing a fuzzy controller, which usually involves only several state variables as the inputs to the controller. This partition strategy needs only a small number of MFs for each input. However, it encounters problems when we have a moderately large number of inputs. For instance, a fuzzy model with 10 inputs and 2 MFs on each input would result in $2^{10} = 1024$ fuzzy if-then rules, which is prohibitively large. This problem, usually referred to as the "curse of dimensionality", can be alleviated by other partition strategies.
- Tree partition: In this method each region can be uniquely specified along a corresponding decision tree. The tree partition relieves the problem of an exponential increase in the number of rules. However, more MFs for each input are needed to define these fuzzy regions, and these MFs do not usually bear clear linguistic meanings. In other words, ortogonality holds roughly in $X \times Y$, but not in either X or Y alone.
- Scatter partition: By covering a subset of the whole input space that characterizes a region of possible occurrence of the input vectors, the scatter partition can also limit the number of rules to a reasonable amount. However, the scatter partition is usually dictated by desired input-output data pairs and thus, in general, orthogonality does not hold in X, Y or $X \times Y$. This makes it hard to estimate the overall mapping directly from the consequent of each rule's output.

2.4 Type-2 Fuzzy Logic Systems

Fuzzy Logic Systems are comprised of rules. Quite often, the knowledge that is used to build these rules is uncertain. Such uncertainty leads to rules whose antecedents or consequents are uncertain, which translates into uncertain antecedent or consequent membership functions (Karnik & Mendel 1998). Type-1 fuzzy systems (like the ones seen in previous sections), whose membership functions are type-1 fuzzy sets, are unable to directly handle such uncertainties. We describe briefly in this section, type-2 fuzzy systems, in which the antecedent or consequent membership functions are type-2 fuzzy sets. Such sets are fuzzy sets whose membership grades themselves are type-1 fuzzy sets; they are very useful in circumstances where it is difficult to determine an exact membership function for a fuzzy set.

2.4.1 Type-2 Fuzzy Sets

The concept of a type-2 fuzzy set, was introduced by Zadeh (1975) as an extension of the concept of an ordinary fuzzy set (henceforth called a "type-1 fuzzy set"). A type-2 fuzzy set is characterized by a fuzzy membership function, i.e., the membership grade for each element of this set is a fuzzy set in [0,1], unlike a type-1 set where the membership grade is a crisp number in [0,1]. Such sets can be used in situations where there is uncertainty about the membership grades themselves, e.g., an uncertainty in the shape of the membership function or in some of its parameters. Consider the transition from ordinary sets to fuzzy sets. When we cannot determine the membership of an element in a set as 0 or 1, we use fuzzy sets of type-1. Similarly, when the situation is so fuzzy that we have trouble determining the membership grade even as a crisp number in [0,1], we use fuzzy sets of type-2.

This does not mean that we need to have extremely fuzzy situations to use type-2 fuzzy sets. There are many real-world problems where we cannot determine the exact form of the membership functions, e.g., in time series prediction because of noise in the data. Another way of viewing this is to consider type-1 fuzzy sets as a first order approximation to the uncertainty in the real-world. Then type-2 fuzzy sets can be considered as a second order approximation. Of course, it is possible to consider fuzzy sets of higher types but the complexity of the fuzzy system increases very rapidly. For this reason, we will only consider very briefly type-2 fuzzy sets. Lets consider some simple examples of type-2 fuzzy sets.

<u>Example</u> 2.1 Consider the case of a fuzzy set characterized by a Gaussian membership function with mean m and a standard deviation that can take values in $[\sigma_1, \sigma_2]$, i.e.,

$$\mu(x)=\exp\{-\tfrac{1}{2}[(x-m)/\sigma]^2\};\ \sigma \in [\sigma_1, \sigma_2] \qquad (2.25)$$

Corresponding to each value of σ, we will get a different membership curve (see Figure 2.8). So, the membership grade of any particular x (except x=m) can take any of a number of possible values depending upon the value of σ, i.e., the membership grade is not a crisp number, it is a fuzzy set. Figure 2.8 shows the domain of the fuzzy set associated with x=0.7; however, the membership function associated with this fuzzy set is not shown in the figure.

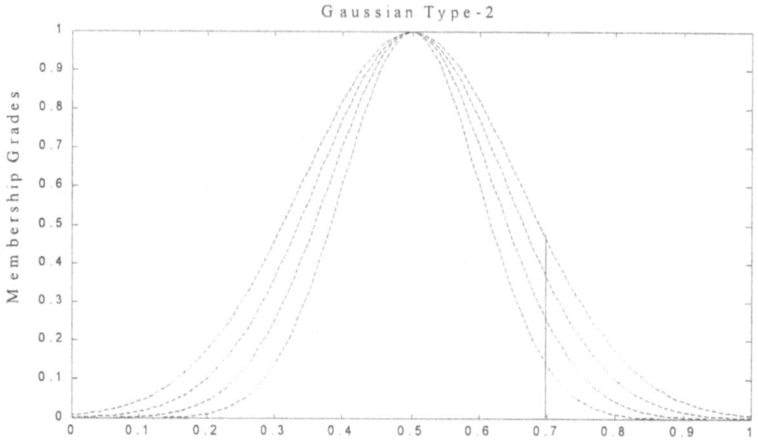

Figure 2.8 A type-2 fuzzy set representing a type-1 fuzzy set with uncertain standard deviation.

Example 2.2 Consider the case of a fuzzy set with a Gaussian membership function having a fixed standard deviation σ, but an uncertain mean, taking values in $[m_1, m_2]$, i.e.,

$$\mu(x)=\exp\{-\tfrac{1}{2}[(x-m)/\sigma]^2\};\ m \in [m_1,m_2] \quad (2.26)$$

Again, μ(x) is a fuzzy set. Figure 2.9 shows an example of such a set.

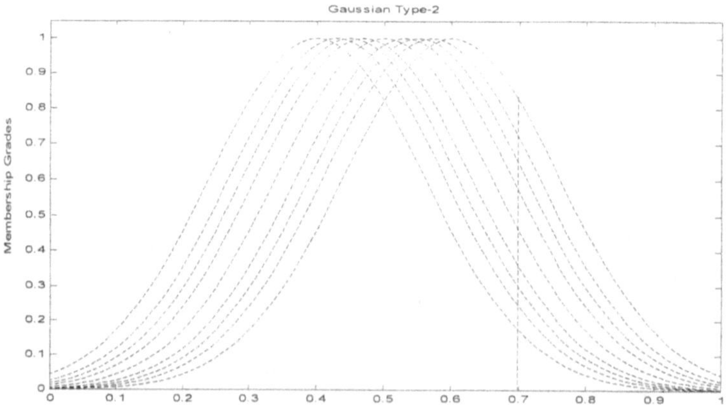

Figure 2.9 A type-2 fuzzy set representing a type-1 fuzzy set with uncertain mean. The mean is uncertain in the interval [0.4, 0.6].

Example 2.3 Consider a type-1 fuzzy set characterized by a Gaussian membership function (mean M and standard deviation σ_x), which gives one crisp membership m(x) for each input $x \in X$, where

$$m(x)=\exp\{-\tfrac{1}{2}[(x-M)/\sigma_x]^2\} \tag{2.27}$$

This is shown in Figure 2.10. Now, imagine that this membership of x is a fuzzy set. Let us call the domain elements of this set "primary memberships" of x (denoted by μ_1) and membership grades of these primary memberships "secondary memberships" of x [denoted by $\mu_2(x, \mu_1)$]. So, for a fixed x, we get a type-1 fuzzy set whose domain elements are primary memberships of x and whose corresponding membership grades are secondary memberships of x. If we assume that the secondary memberships follow a Gaussian with mean m(x) and standard deviation σ_m, as in Figure 2.10, we can describe the secondary membership function for each x as

$$\mu_2(x,\mu_1) = e - \tfrac{1}{2}[(\mu_1 - m(x))/\sigma_m]^2 \tag{2.28}$$

where $\mu_1 \in [0,1]$ and m is as in equation (2.27).

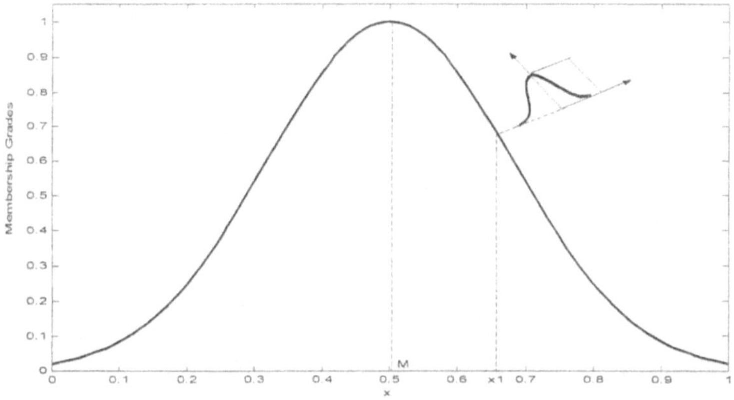

Figure 2.10 A type-2 fuzzy set in which the membership grade of every domain point is a Gaussian type-1 set.

We can formally define these two kinds of type-2 sets as follows.

Definition 2.20 **Gaussian type-2**
A Gaussian type-2 fuzzy set is one in which the membership grade of every domain point is a Gaussian type-1 set contained in [0,1].

Example 2.3 shows an example of a Gaussian type-2 fuzzy set.

Definition 2.21 **Interval type-2**
An interval type-2 fuzzy set is one in which the membership grade of every domain point is a crisp set whose domain is some interval contained in [0,1].

Example 2.1 shows an example of an interval type-2 fuzzy set.

2.4.2 Type-2 Fuzzy Systems

The basics of fuzzy logic do not change from type-1 to type-2 fuzzy sets, and in general, will not change for any type-n (Karnik & Mendel 1998). A higher-type number just indicates a higher "degree of fuzziness". Since a higher type changes the nature of the membership functions, the operations that depend on the membership functions change; however, the basic principles of fuzzy logic are independent of the nature of membership functions and hence, do not change. Rules of inference like Generalized Modus Ponens or Generalized Modus Tollens continue to apply.

In Figure 2.11 we show the general structure of a type-2 fuzzy system. We assume that both antecedent and consequent sets are type-2; however, this need not necessarily be the case in practice.

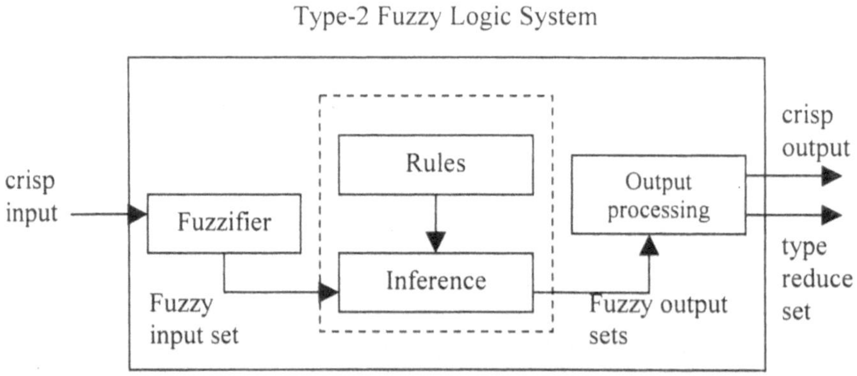

Figure 2.11 General structure of a type-2 fuzzy system. The structure of the "output processing" block is shown in Figure (b). In order to emphasize the importance of the type-reduced set, we have shown two outputs for the fuzzy system, the type reduced set and the crisp defuzzified value.

The structure of the type-2 fuzzy rules is the same as for the type-1 case because the distinction between type-2 and type-1 is associated with the nature of the membership functions. Hence, the only difference is that now some or all the sets involved in the rules are of type-2. In a type-1 fuzzy system, where the output sets are type-1 fuzzy sets, we perform defuzzification in order to get a number, which is in some sense a crisp (type-0) representative of the combined output sets. In the type-2 case, the output sets are type-2; so we have to use extended versions of type-1 defuzzification methods. Since type-1 defuzzification gives a crisp number at the output of the fuzzy system, the extended defuzzification operation in the type-2 case gives a type-1 fuzzy set at the output. Since this operation takes us from the type-2 output sets of the fuzzy system to a type-1 set, we can call this operation "type reduction" and call the type-1 fuzzy set so obtained a "type-reduced set". The type-reduced fuzzy set may then be defuzzified to obtain a single crisp number; however, in many applications, the type-reduced set may be more important than a single crisp number.

Type-2 sets can be used to convey the uncertainties in membership functions of type-1 fuzzy sets, due to the dependence of the membership functions on available linguistic and numerical information. Linguistic information (e.g. rules from experts), in general, does not give any information about the shapes of the membership functions. When membership functions are determined or tuned based on numerical data, the uncertainty in the numerical data, e.g., noise, translates into uncertainty in the membership functions. In all such cases, any available information about the linguistic/numerical uncertainty can be incorporated in the type-2 framework. However, even with all of the advantages that fuzzy type-2 systems have, the literature on the applications of type-2 sets is scarce. Some examples are Yager (1980) for decision making, and Wagenknecht & Hartmann (1988) for solving fuzzy relational equations. We think that more applications of type-2 fuzzy systems will come in the near future as the area matures and the theoretical results become more understandable for the general public in the fuzzy arena.

2.5 Fuzzy Modelling

In general, we design a fuzzy inference system based on the past known behavior of a target system. The fuzzy system is then expected to be able to reproduce the behavior of the target system. For example, if the target system is a human operator in charge of a chemical reaction process, then the fuzzy inference system becomes a fuzzy logic controller that can regulate and control the process.

Let us now consider how we might construct a fuzzy inference system for a specific application. Generally speaking, the standard method for constructing a fuzzy inference system, a process usually called "fuzzy modelling", has the following features:

- The rule structure of a fuzzy inference system makes it easy to incorporate human expertise about the target system directly into the modelling process. Namely, fuzzy modelling takes advantage of "domain knowledge" that might not be easily or directly employed in other modelling approaches.
- When the input-output data of a target system is available, conventional system identification techniques can be used for fuzzy modelling. In other words, the use of "numerical data" also plays an important role in "fuzzy modelling", just as in other mathematical modelling methods.

Conceptually, fuzzy modelling can be pursued in two stages, which are not totally disjoint. The first stage is the identification of the "surface structure", which includes the following tasks:

1. Select relevant input and output variables.
2. Choose a specific type of fuzzy inference system.
3. Determine the number of linguistic terms associated with each input and output variables.
4. Design a collection of fuzzy if-then rules.

Note that to accomplish the preceding tasks, we rely on our own knowledge (common sense, simple physical laws, an so on) of the target system, information provided by human experts who are familiar with the target system, or simply trial and error.

After the first stage of fuzzy modelling, we obtain a rule base that can more or less describe the behavior of the target system by means of linguistic terms. The meaning of these linguistic terms is determined in the second stage, the identification of "deep structure", which determines the MFs of each linguistic term. Specifically, the identification of deep structure includes the following tasks:

1. Choose an appropriate family of parameterized MFs.
2. Interview human experts familiar with the target systems to determine the parameters of the MFs used in the rule base.
3. Refine the parameters of the MFs using regression and optimization techniques.

Task 1 and 2 assume the availability of human experts, while task 3 assumes the availability of a desired input-output data set.

2.6 Summary

In this chapter, we have presented the main ideas underlying Fuzzy Logic and we have only started to point out the many possible applications of this powerful computational theory. We have discussed in some detail fuzzy set theory, fuzzy reasoning and fuzzy inference systems. We also reviewed briefly type-2 fuzzy logic, which is a generalization of conventional fuzzy logic (type-1). At the end, we also gave some remarks about fuzzy modelling. In the following chapters, we will show how fuzzy logic techniques (in some cases, in conjunction with other methodologies) can be applied to solve real world complex problems. This chapter will serve as a basis for the new hybrid intelligent methods, for modelling and simulation, that will be described in Chapters 6, 7 and 8 of this book. Fuzzy Logic will also play an important role in the new neuro-fuzzy methodology for control that is presented in Chapters 9 to 13 of this book.

Chapter 3

Neural Networks for Control

Application of fuzzy inference systems to automatic control was first reported in Mamdani's paper (Mamdani & Assilian, 1975), where a "fuzzy logic controller" (FLC) was used to emulate a human operator's control of a steam engine and boiler combination. Since then, "fuzzy logic control" has been recognized as the most significant and fruitful application for fuzzy logic (Kosko, 1992). In the past few years, advances in microprocessors and hardware technologies have created an even more diversified application domain for fuzzy logic controllers, which ranges from consumer electronics to the automobile industry. However, without adaptive capability, the performance of FLCs relies exclusively on two factors: the availability of human experts, and the knowledge acquisition techniques to convert human expertise into appropriate fuzzy rules. These two factors substantially restrict the application domain of FLCs.

On the other hand, investigation into using neural networks in automatic control systems did not receive much attention until the "backpropagation" learning rule was formulated by Rumelhart and others (Rumelhart, Hinton & Williams, 1986). Since then, research of neural control has evolved quickly and a number of neural controller design methods have been proposed in the literature (Werbos, 1991).

Figure 3.1 is a block diagram of a typical "feedback control system", where the "plant" (or "process") represents the dynamic system to be controlled and the "controller" employs a control strategy to achieve a control goal. Here we shall denote the state variables of the plant as a vector x(t); these variables are usually governed by a set of "state equations" (usually differential equations) that characterize the dynamic behavior of the plant. Since the state variables are internal to the plant, some of them may not be directly measurable from the external world. The measurable quantities of the plant, also known as its outputs, are denoted as a vector y(t). We shall assume that all states are measurable; thus the output of the plant y(t) is equal to the state x(t).

The state equation for a general non-linear plant can be expressed in the matrix notation

$$\mathbf{x}'(t) = \mathbf{f}\,(\mathbf{x}(t), \mathbf{u}(t)) \qquad \text{(plant dynamics)} \qquad (3.1)$$

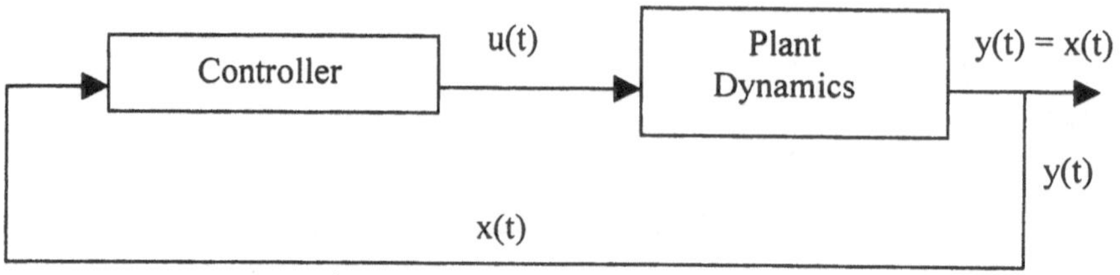

Figure 3.1 Block diagram for a continuous feedback control system.

where u(t) is the controller's output at time t, and the size of the vector $\mathbf{x}(t)$ is called the "order" of the plant. A general control goal is to find a controller with a static function ϕ that maps an observed plant output $\mathbf{x}(t)$ to a control action $\mathbf{u}$-that is, $\mathbf{u}(t) = \phi\,(\mathbf{x}(t))$- such that the plant output can follow some given desired output signal $\mathbf{x}_d(t)$ as closely as possible. If $\mathbf{x}_d(t)$ is a constant vector, then the control problem is referred to as "regulator problem", where the plant states are directly fed back to the controller. This is actually what Figure 3.1 shows. On the other hand, if the desired trajectory $\mathbf{x}_d(t)$ is a time-varying signal, then we have a "tracking problem" in which an error signal, defined as the difference between desired and actual outputs, is fed back to the controller. If $\mathbf{f}$ is unknown, we need to perform system identification first to find a model for the plant. Moreover, if $\mathbf{f}$ is time varying, it is desirable to make ϕ adaptive to respond to the changing characteristics of the plant.

In the case of linear feedback control systems, the plant and controller can be reformulated as the following equations:

$$\begin{aligned} \mathbf{x}'(t) &= \mathbf{A}\mathbf{x}(t) + \mathbf{B}\mathbf{u}(t) \qquad && \text{(plant dynamics)} \qquad (3.2)\\ \mathbf{u}(t) &= \mathbf{k}\mathbf{x}(t) && \text{(linear controller)} \end{aligned}$$

The treatment of linear control systems is relatively complete in the literature (for example, see Brogan, 1991) and will not be discussed here. On the other hand, the area of non-linear control is still with many open problems and its more interesting. In this book, the treatment will be restricted to non-linear plants with a general form given by Equation (3.1).

If we replace the controller block in Figure 3.1 with neural networks or fuzzy systems, then we end up with "neural" or "fuzzy control systems", respectively. In other words, neural or fuzzy control design methods are systematic ways of constructing neural networks or fuzzy inference systems, respectively, as controllers intended to achieve prescribed control goals. In the same vein, the term "neuro-fuzzy control" has been used when one is speaking about design methods for fuzzy logic controllers that use neural network techniques.

Most neural or fuzzy controllers are nonlinear; thus rigorous analysis for neuro-fuzzy control systems is difficult and remains a challenging area for further investigation. On the other hand, a neuro-fuzzy controller usually contains a large number of parameters; it is thus more versatile than a linear controller in dealing with non-linear plant characteristics. Therefore, neuro-fuzzy controllers almost always surpass pure linear controllers if designed properly.

In this chapter, we present the basic concepts, notation, and basic learning algorithms for neural networks that will be needed in the following chapters of this book. The chapter is organized as follows: Backpropagation for Feedforward Networks, Adaptive Neuro-Fuzzy Inference Systems, Neuro-Fuzzy Control and Adaptive Neuro-Control. First, we give a brief review of the basic concepts of neural networks and the backpropagation learning algorithm. Second, we give a brief description of adaptive neuro-fuzzy systems. Third, we give a brief review on the current methods for neuro-fuzzy control. Finally, we end the chapter with some remarks about adaptive control and model-based control. We consider this material necessary to understand the new methods for control that will be presented in Chapter 7 of this book.

3.1 Backpropagation for Feedforward Networks

This section describes the architectures and learning algorithms for adaptive networks, a unifying framework that subsumes almost all kinds of neural network paradigms with supervised learning capabilities. An adaptive network, as the name indicates, is a network structure consisting of a number of nodes connected through directional links. Each node represents a process unit, and the links between nodes specify the causal relationship between the connected nodes. The learning rule specifies how the parameters (of the nodes) should be updated to minimize a prescribed error measure.

The basic learning rule of the adaptive network is the well-known steepest descent method, in which the gradient vector is derived by successive invocations of the chain rule. This method for systematic calculation of the gradient vector was proposed independently several times, by Bryson and Ho (1969), Werbos (1974), and Parker (1982). However, because research on artificial neural networks was still in its infancy at those times, these researchers' early work failed to receive the attention it deserved. In 1986, Rumelhart et al. used the same procedure to find the gradient in a multilayer neural network. Their procedure was called "backpropagation learning rule", a name which is now widely known because the work of Rumelhart inspired enormous interest in research on neural networks. In this section, we introduce Werbos's original backpropagation method for finding gradient vectors and also present improved versions of this method.

3.1.1 The Backpropagation Learning Algorithm

Suppose that a given feedforward adaptive network has L layers and layer l (l = 0, 1,..., L) has N(l) nodes. Then the output and function of node i [i = 1, ..., N(l)] in layer l can be represented as $x_{l,i}$ and $f_{l,i}$, respectively, as shown in Figure 3.2. Since the output of a node depends on the incoming signals and the parameter set of the node, we have the following general expression for the node function $f_{l,i}$:

$$x_{l,i} = f_{l,i}(x_{l-1,1}, \ldots, x_{l-1,N(l-1)}, \alpha, \beta, \gamma, \ldots) \tag{3.3}$$

where α, β, γ, etc. are the parameters of this node.

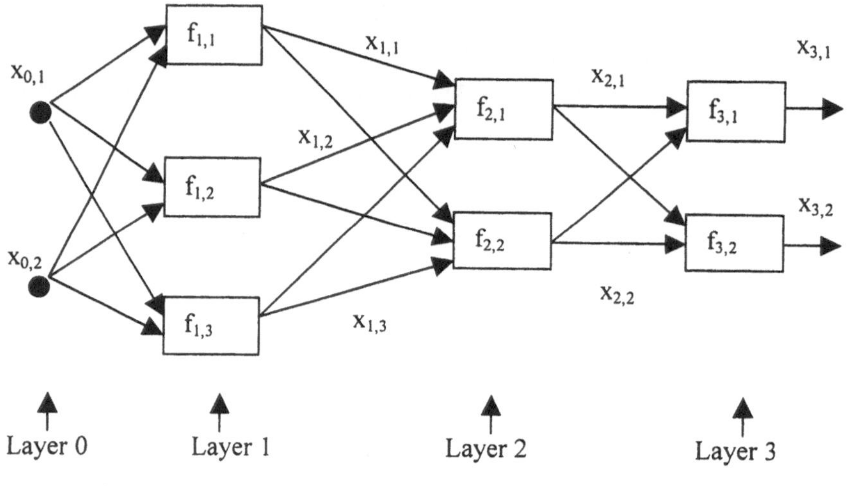

Figure 3.2 Feedforward adaptive network.

Assuming that the given training data set has P entries, we can define an error measure for the pth ($1 \leq p \leq P$) entry of the training data as the sum of the squared errors:

$$E_p = \sum_{k=1}^{N(L)} (d_k - x_{L,k})^2 \tag{3.4}$$

where d_k is the kth component of the pth desired output vector and $x_{L,k}$ is the kth component of the actual output vector produced by presenting the pth input vector to the network. Obviously, when E_p is equal to zero, the network is able to reproduce exactly the desired output vector in the pth training data pair. Thus our task here is to minimize an overall error measure, which is defined as $E = \sum E_p$.

We can also define the "error signal" $\varepsilon_{l,i}$ as the derivative of the error measure E_p with respect to the output of the node i in layer l, taking both direct and indirect paths into consideration. Mathematically,

$$\varepsilon_{l,i} = \frac{\partial^+ E_p}{\partial x_{l,i}} \tag{3.5}$$

this expression was called the "ordered derivative" by Werbos (1974). The difference between the ordered derivative and the ordinary partial derivative lies in the way we view the function to be differentiated. For an internal node output $x_{l,i}$, the partial derivative $\partial^+ E_p / \partial x_{l,i}$ is equal to zero, since E_p does not depend on $x_{l,i}$ directly. However, it is obvious that E_p does depend on $x_{l,i}$ indirectly, since a change in $x_{l,i}$ will propagate through indirect paths to the output layer and thus produce a corresponding change in the value of E_p.

The error signal for the ith output node (at layer L) can be calculated directly:

$$\varepsilon_{L,i} = \frac{\partial^+ E_p}{\partial x_{L,i}} = \frac{\partial E_p}{\partial x_{L,i}} \tag{3.6}$$

This is equal to $\varepsilon_{L,i} = -2(d_i - x_{L,i})$ if E_p is defined as in Equation (3.4). For the internal node at the ith position of layer l, the error signal can be derived by the chain rule of differential calculus:

$$\underbrace{\varepsilon_{l,i} = \frac{\partial^+ E_p}{\partial x_{l,i}}}_{\text{error signal at layer l}} = \sum_{m=1}^{N(l+1)} \underbrace{\frac{\partial^+ E_p}{\partial x_{l+1,m}}}_{\text{error signal at layer l+1}} \frac{\partial f_{l+1,m}}{\partial x_{l,i}} = \sum_{m=1}^{N(l+1)} \varepsilon_{l+1,m} \frac{\partial f_{l+1,m}}{\partial x_{l,i}} \tag{3.7}$$

where $0 \leq l \leq L-1$. That is, the error signal of an internal node at layer l can be expressed as a linear combination of the error signal of the nodes at layer l+1.

Therefore, for any l and i, we can find $\varepsilon_{l,i}$ by first applying Equation (3.6) once to get error signals at the output layer, and then applying Equation (3.7) iteratively until we reach the desired layer l. The underlying procedure is called backpropagation since the error signals are obtained sequentially from the output layer back to the input layer.

The gradient vector is defined as the derivative of the error measure with respect to each parameter, so we have to apply the chain rule again to find the gradient vector. If α is a parameter of the ith node at layer l, we have

$$\frac{\partial^+ E_p}{\partial \alpha} = \frac{\partial^+ E_p}{\partial x_{l,i}} \frac{\partial f_{l,i}}{\partial \alpha} = \varepsilon_{l,i} \frac{\partial f_{l,i}}{\partial \alpha} \tag{3.8}$$

The derivative of the overall error measure E with respect to α is

$$\frac{\partial^+ E}{\partial \alpha} = \sum_{p=1}^{P} \frac{\partial^+ E_p}{\partial \alpha} \tag{3.9}$$

Accordingly, for simple steepest descent (for minimization), the update formula for the generic parameter α is

$$\Delta\alpha = -\eta \frac{\partial^+ E}{\partial \alpha} \tag{3.10}$$

in which η is the "learning rate", which can be further expressed as

$$\eta = \frac{k}{\sqrt{\Sigma_\alpha (\partial E / \partial \alpha)^2}} \tag{3.11}$$

where k is the "step size", the length of each transition along the gradient direction in the parameter space.

There are two types of learning paradigms that are available to suit the needs for various applications. In "off-line learning" (or "batch learning"), the update formula for parameter α is based on Equation (3.9) and the update action takes place only after the whole training data set has been presented-that is, only after each "epoch" or "sweep". On the other hand, in "on-line learning" (or "pattern-by-pattern learning"), the parameters are updated immediately after each input-output pair has been presented, and the update formula is based on Equation (3.8). In practice, it is possible to combine these two learning modes and update the parameter after k training data entries have been presented, where k is between 1 and P and it is sometimes referred to as the "epoch size".

3.1.2 Backpropagation Multilayer Perceptions

Artificial neural networks, or simply "neural networks" (NNs), have been studied for more than three decades since Rosenblatt first applied single-layer "perceptrons" to pattern classification learning (Rosenblatt, 1962). However, because Minsky and Papert pointed out that single-layer systems were limited and expressed pessimism over multilayer systems, interest in NNs dwindled in the 1970s (Minsky & Papert, 1969). The recent resurgence of interest in the field of NNs has been inspired by new developments in NN learning algorithms (Fahlman & Lebiere, 1990), analog VLSI circuits, and parallel processing techniques (Lippmann, 1987).

Quite a few NN models have been proposed and investigated in recent years. These NN models can be classified according to various criteria, such as their learning methods (supervised versus unsupervised), architectures (feedforward versus recurrent), output types (binary versus continuous), and so on. In this section, we confine our scope to modelling problems with desired input-output data sets, so the resulting networks must have adjustable parameters that are updated by a supervised learning rule. Such networks are often referred to as "supervised learning" or "mapping networks", since we are interested in shaping the input-output mappings of the network according to a given training data set.

A backpropagation "multilayer perceptron" (MLP) is an adaptive network whose nodes (or neurons) perform the same function on incoming signals; this node function is usually a composite of the weighted sum and a differentiable non-linear activation function, also known as the "transfer function". Figure 3.3 depicts three of the most commonly used activation functions in backpropagation MLPs:

Logistic function: $$f(x) = \frac{1}{1 + e^{-x}}$$

Hyperbolic tangent function: $$f(x) = \tan h\,(x/2) = \frac{1 - e^{-x}}{1 + e^{-x}}$$

Identity function: $$f(x) = x$$

Both the hyperbolic tangent and logistic functions approximate the signum and step function, respectively, and provide smooth, nonzero derivatives with respect to input signals. Sometimes these two activation functions are referred to as "squashing functions" since the inputs to these functions are squashed to the range [0,1] or [-1,1]. They are also called "sigmoidal functions" because their s-shaped curves exhibit smoothness and asymptotic properties.

Backpropagation MLPs are by far the most commonly used NN structures for applications in a wide range of areas, such as pattern recognition, signal processing, data compression and automatic control. Some of the well-known instances of applications include NETtalk (Sejnowski & Rosenberg, 1987), which trained an MLP to pronounce English text, Carnegie Mellon University's ALVINN (Pomerleau, 1991), which used an MLP for steering an autonomous vehicle; and optical character recognition (Sakinger, Boser, Bromley, Lecun & Jackel, 1992). In the following lines, we derive the backpropagation learning rule for MLPs using the logistic function.

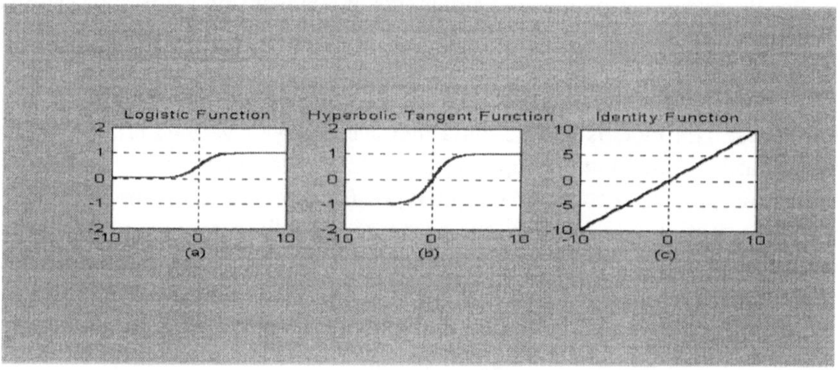

Figure 3.3 Activation functions for backpropagation MLPs: (a) logistic function; (b) hyperbolic function; (c) identity function.

The "net input" $\bar{x}$ of a node is defined as the weighted sum of the incoming signals plus a bias term. For instance, the net input and output of node j in Figure 3.4 are

$$\bar{x}_j = \sum_i w_{ij}\, x_i + w_j \,, \qquad (3.12)$$
$$x_j = f(\bar{x}_j) = \frac{1}{1 + \exp(-\bar{x}_j)} \,,$$

where x_i is the output of node i located in any one of the previous layers, w_{ij} is the weight associated with the link connecting nodes i and j, and w_j is the bias of node j. Since the weights w_{ij} are actually internal parameters associated with each node j, changing the weights of a node will alter the behavior of the node and in turn alter the behavior of the whole backpropagation MLP.

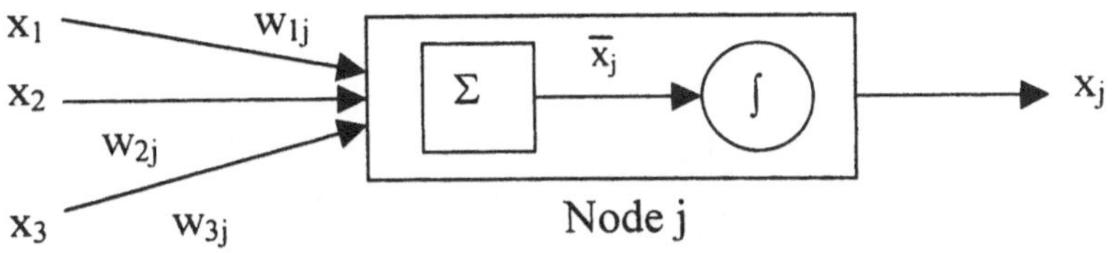

Figure 3.4 Node j of a backpropagation MLP.

Figure 3.5 shows a three-layer backpropagation MLP with three inputs to the input layer, three neurons in the hidden layer, and two output neurons in the output layer. For simplicity, this MLP will be referred to as a 3-3-2 network, corresponding to the number of nodes in each layer.

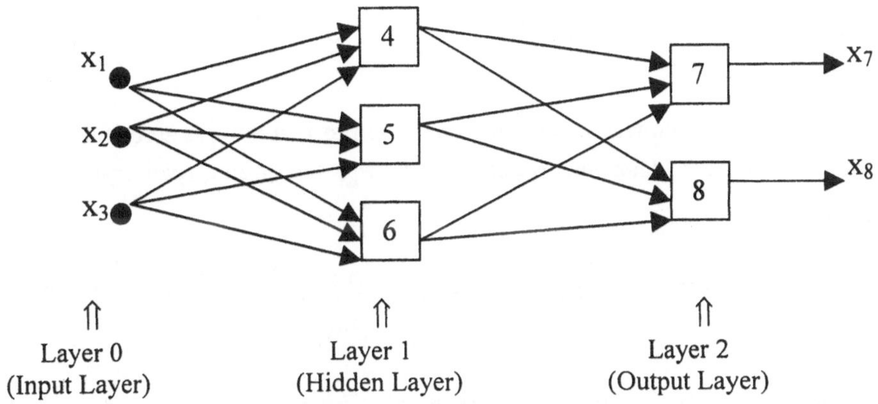

Figure 3.5 A 3-3-2 backpropagation MLP.

The "backward error propagation", also known as the "backpropagation" (BP) or the "generalized data rule" (GDR), is explained next. First, a squared error measure for the pth input-output pair is defined as

$$E_p = \sum_k (d_k - x_k)^2 \quad (3.13)$$

where d_k is the desired output for node k, and x_k is the actual output for node k when the input part of the pth data pair presented. To find the gradient vector, an error term $\bar{\varepsilon}_i$ is defined as

$$\bar{\varepsilon}_i = \frac{\partial^+ E_p}{\partial x_i} \qquad (3.14)$$

By the chain rule, the recursive formula for ε_i can be written as

$$\bar{\varepsilon}_i = \begin{cases} -2(d_i - x_i) \dfrac{\partial x_i}{\partial \bar{x}_i} = -2(d_i - x_i)\, x_i\,(1 - x_i) & \text{if node i is a output node} \\ \dfrac{\partial x_i}{\partial \bar{x}_i} \sum_{j,i<j} \dfrac{\partial^+ E_p}{\partial \bar{x}_j} \dfrac{\partial \bar{x}_j}{\partial x_i} = x_i\,(1 - x_i) \sum_{j,i<j} \bar{\varepsilon}_j\, w_{ij} & \text{otherwise} \end{cases} \qquad (3.15)$$

where w_{ij} is the connection weight from node i to j; and w_{ij} is zero if there is no direct connection. Then the weight update w_{ki} for on-line (pattern-by-pattern) learning is

$$\Delta w_{ki} = -\eta \frac{\partial^+ E_p}{\partial w_{ki}} = -\eta \frac{\partial^+ E_p}{\partial \bar{x}_i} \frac{\partial \bar{x}_i}{\partial w_{ki}} = -\eta\, \bar{\varepsilon}_i\, x_k \qquad (3.16)$$

where η is a learning rate that affects the convergence speed and stability of the weights during learning.

For off-line (batch) learning, the connection weight w_{ki} is updated only after presentation of the entire data set, or only after an "epoch":

$$\Delta w_{ki} = -\eta \frac{\partial^+ E}{\partial w_{ki}} = -\eta \sum_p \frac{\partial^+ E_p}{\partial w_{ki}} \qquad (3.17)$$

or, in vector form,

$$\Delta \mathbf{w} = - \eta \frac{\partial^{+} E}{\partial \mathbf{w}} = - \eta \nabla_{\mathbf{w}} E \tag{3.18}$$

where $E = \Sigma_p E_p$. This corresponds to a way of using the true gradient direction based on the entire data set.

The approximation power of backpropagation MLPs has been explored by some researchers. Yet there is very little theoretical guidance for determining network size in terms of say, the number of hidden nodes and hidden layers it should contain. Cybenko (1989) showed that a backpropagation MLP, with one hidden layer and any fixed continuous sigmoidal non-linear function, can approximate any continuous function arbitrarily well on a compact set. When used as a binary-valued neural network with the step activation function, a backpropagation MLP with two hidden layers can form arbitrary complex decision regions to separate different classes, as Lippmann (1987) pointed out. For function approximation as well as data classification, two hidden layers may be required to learn a piecewise-continuous function (Masters, 1993).

3.2 Adaptive Neuro-Fuzzy Inference Systems

In this section, we describe a class of adaptive networks that are functionally equivalent to fuzzy inference systems (Kosko, 1992). The architecture is referred to as ANFIS, which stands for "adaptive network-based fuzzy inference system". We describe how to decompose the parameter set to facilitate the hybrid learning rule for ANFIS architectures representing both the Sugeno and Tsukamoto fuzzy models.

3.2.1 ANFIS Architecture

A fuzzy inference system consists of three conceptual components: a fuzzy rule base, which contains a set of fuzzy if-then rules; a database, which defines the membership functions used in the fuzzy rules; and a reasoning mechanism, which performs the inference procedure upon the rules to derive a reasonable output or conclusion (Kandel, 1992). For simplicity, we assume that the fuzzy inference system under consideration has two inputs x and y and one output z. For a first-order Sugeno fuzzy model (Sugeno & Kang, 1988), a common rule set with two fuzzy if-then rules is the following:

Rule 1: If x is A_1 and y is B_1, then $f_1 = p_1 x + q_1 y + r_1$,

Rule 2: If x is A_2 and y is B_2, then $f_2 = p_2 x + q_2 y + r_2$,

Figure 3.6 (a) illustrates the reasoning mechanism for this Sugeno model; the corresponding equivalent ANFIS architecture is as shown in Figure 3.6 (b), where nodes of the same layer have similar functions, as described next. (Here we denote the output of the ith node in layer l as $0_{l,i}$).

Layer 1: Every node i in this layer is an adaptive node with a node function

$$0_{l,i} = \mu_{Ai}(x), \text{ for } i = 1, 2,$$
$$0_{l,i} = \mu_{Bi-2}(y), \text{ for } i = 3, 4, \qquad (3.19)$$

where x (or y) is the input to node i and A_i (or B_{i-2}) is a linguistic label (such as "small" or "large") associated with this node. In other words, $0_{l,i}$ is the membership grade of a fuzzy set A and it specifies the degree to which the given input x (or y) satisfies the quantifier A. Here the membership function for A can be any appropriate parameterized membership function, such as the generalized bell function:

$$\mu_A(x) = \frac{1}{1 + |(x-c_i)/a_i|^{2bi}} \qquad (3.20)$$

where $\{a_i, b_i, c_i\}$ is the parameter set. As the values of these parameters change, the bell-shaped function varies accordingly, thus exhibiting various forms of membership functions for a fuzzy set A. Parameters in this layer are referred to as "premise parameters".

Layer 2: Every node in this layer is a fixed node labeled Π, whose output is the product of all incoming signals:

$$0_{2,i} = w_i = \mu_{Ai}(x)\, \mu_{Bi}(y), \quad i = 1, 2. \qquad (3.21)$$

Each node output represents the firing strength of a fuzzy rule.

Layer 3: Every node in this layer is a fixed node labeled N. The ith node calculates the ratio of the ith rule's firing strength to the sum of all rules' firing strengths:

$$0_{3,i} = \overline{w_i} = w_i / (w_1 + w_2), \quad i = 1,2. \qquad (3.22)$$

For convenience, outputs of this layer are called "normalized firing strengths".

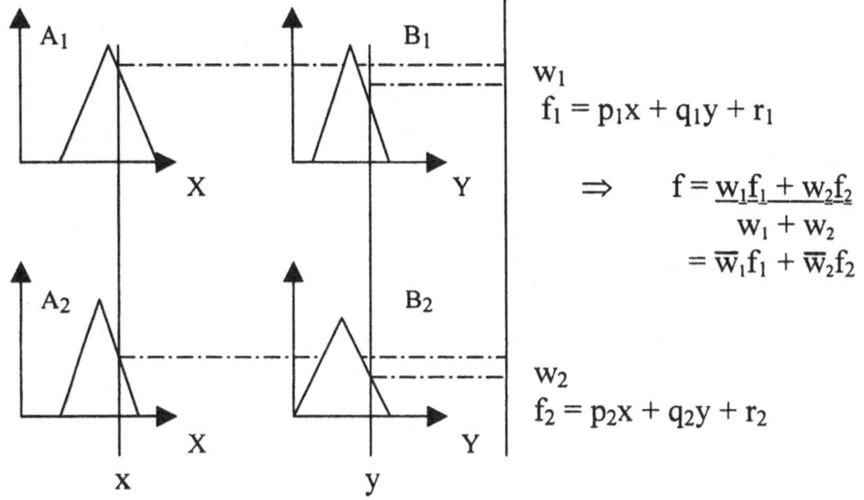

(a)

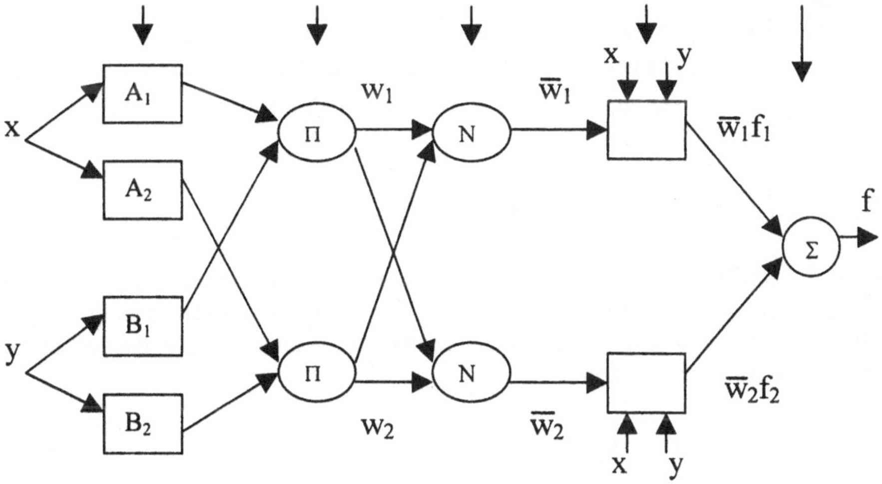

(b)

Figure 3.6 (a) A two-input Sugeno fuzzy model with 2 rules; (b) equivalent ANFIS architecture (adaptive nodes shown with a square and fixed nodes with a circle).

Layer 4: Every node i in this layer is an adaptive node with a node function

$$O_{4,i} = \overline{w}_i f_i = \overline{w}_i (p_i x + q_i y + r_i) \quad , \tag{3.23}$$

where $\overline{w}_i$ is a normalized firing strength from layer 3 and $\{p_i , q_i , r_i \}$ is the parameter set of this node. Parameters in this layer are referred to as "consequent parameters".

Layer 5: The single node in this layer is a fixed node labeled Σ, which computes the overall output as the summation of all incoming signals:

$$\text{overall output} = O_{5,i} = \sum_i \overline{w}_i f_i = \frac{\sum_i w_i f_i}{\sum_i w_i} \tag{3.24}$$

Thus we have constructed an adaptive network that is functionally equivalent to a Sugeno fuzzy model. We can note that the structure of this adaptive network is not unique; we can combine layers 3 and 4 to obtain an equivalent network with only four layers. In the extreme case, we can even shrink the whole network into a single adaptive node with the same parameter set. Obviously, the assignment of node functions and the network configuration are arbitrary, as long as each node and each layer perform meaningful and modular functionalities.

The extension from Sugeno ANFIS to Tsukamoto ANFIS is straightforward, as shown in Figure 3.7, where the output of each rule (f_i, i = 1, 2) is induced jointly by a consequent membership function and a firing strength.

3.2.2 Learning Algorithm

From the ANFIS architecture shown in Figure 3.6 (b), we observe that when the values of the premise parameters are fixed, the overall output can be expressed as a linear combination of the consequent parameters. Mathematically, the output f in Figure 3.6 (b) can be written as

$$f = \frac{w_1}{w_1 + w_2} f_1 + \frac{w_2}{w_1 + w_2} f_2 \tag{3.25}$$

$$= \overline{w}_1 (p_1 x + q_1 y + r_1) + \overline{w}_2 (p_2 x + q_2 y + r_2)$$

$$= (\overline{w}_1 x) p_1 + (\overline{w}_1 y) q_1 + (\overline{w}_1) r_1 + (\overline{w}_2 x) p_2 + (\overline{w}_2 y) q_2 + (\overline{w}_2) r_2$$

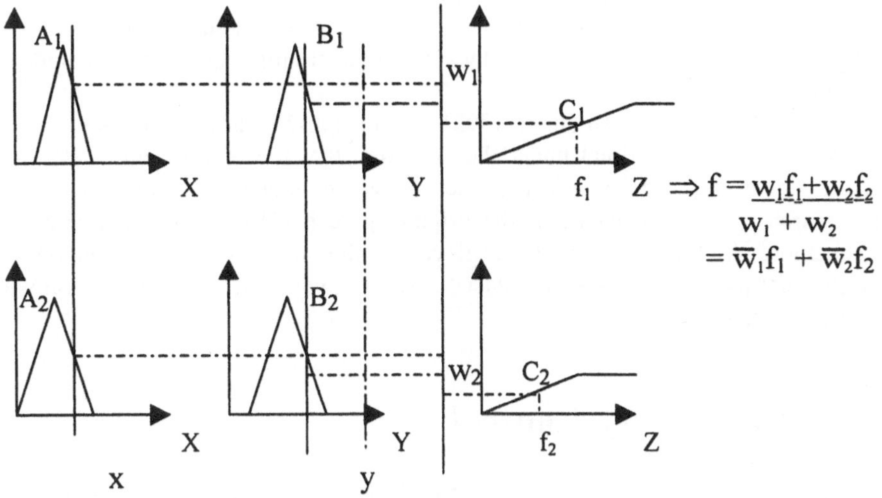

(a)

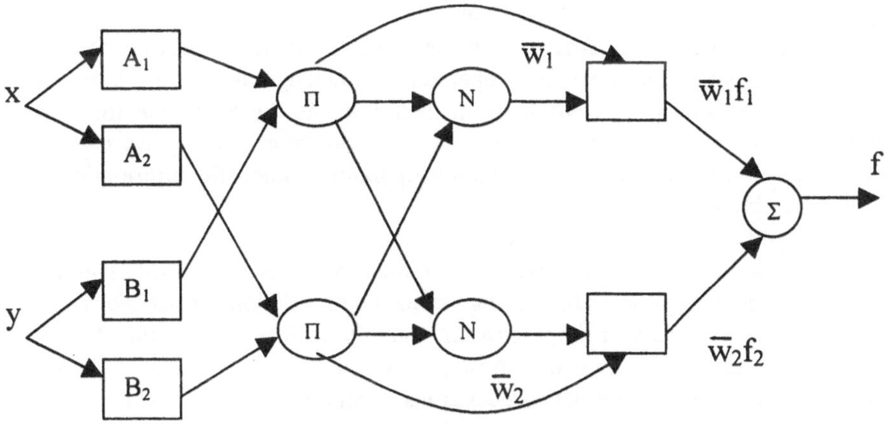

(b)

Figure 3.7 (a) A two-input Tsukamoto fuzzy model with two rules; (b) equivalent ANFIS architecture.

which is linear in the consequent parameters p_1, q_1, r_1, p_2, q_2, and r_2. From this observation, we can use a hybrid learning algorithm for parameter estimation in this kind of models (Jang, 1993). More specifically, in the forward pass of the hybrid learning algorithm, node outputs go forward until layer 4 and the consequent parameters are identified by the least-squares method. In the backward pass, the error signals propagate backward and the premise parameters are updated by gradient descent.

It has been shown (Jang, 1993) that the consequent parameters identified in this manner are optimal under the condition that the premise parameters are fixed. Accordingly, the hybrid approach converges much faster since it reduces the search space dimensions of the original pure backpropagation method. For Tsukamoto ANFIS, this can be achieved if the membership function on the consequent part of each rule is replaced by a piecewise linear approximation with two consequent parameters.

3.3 Neuro-Fuzzy Control

The original purpose of fuzzy logic control, as proposed in Mamdani's paper in 1975, was to mimic the behavior of a human operator able to control a complex plant satisfactorily. The complex plant in question could be a chemical reaction process, a subway train, or a traffic signal control system. After more than 20 years, the ultimate goal of fuzzy controllers remains the same-that is, to automate an entire control process by replacing a human operator with a fuzzy controller made up of computer software/hardware.

To construct a fuzzy controller, we need to perform "knowledge acquisition", which takes a human operator's knowledge about how to control a system and generates a set of fuzzy if-then rules as the backbone for a fuzzy controller that behaves like the original human operator. Usually we can obtain two types of information from a human operator: "linguistic information" and "numerical information".

Linguistic information: An experienced human operator can usually summarize his or her reasoning process in arriving at final control actions or decisions as a set of fuzzy if-then rules with imprecise but roughly correct membership functions; this corresponds to the linguistic information supplied by human experts, which is obtained via a lengthy interview process plus a certain amount of trial and error.

Numerical information: When a human operator is working, it is possible to record the sensor data observed by the human and the human's corresponding actions as a set of desired input-output data pairs. This data set can be used as training data in constructing a fuzzy controller.

Prior to the emergence of neuro-fuzzy approaches, most design methods used only linguistic information to build fuzzy controllers; this approach is not easily formalized and is more of an art than an engineering practice. Following this approach usually involves manual trial-and-error processes to fine-tune the membership functions. Successful fuzzy control applications based on linguistic information plus trial-and-error tuning include steam engine and boiler control (Mamdani & Assilian, 1975), Sendai subway systems (Yasunobu & Miyamoto, 1985), nuclear reaction control (Bernard, 1988), automobile transmission control (Kasai & Morimoto, 1988), aircraft control (Chiu, Chand, Moore & Chaudhary, 1991), and many others.

Now, with learning algorithms, we can take further advantage of the numerical information (input-output data pairs) and refine the membership functions in a systematic way. In other words, we can use linguistic information to identify the structure of a fuzzy controller, and then use numerical information to identify the parameters such that the fuzzy controller can reproduce the desired action more accurately.

3.3.1 Inverse Learning

The development of "inverse learning" (Widrow & Stearns, 1985) for designing neuro-fuzzy controllers involves two phases. In the learning phase, an on-line or off-line technique is used to model the inverse dynamics of the plant. The obtained neuro-fuzzy model, which represents the inverse dynamics of the plant, is then used to generate control actions in the application phase. These two phases, can proceed simultaneously, hence this design method fits in perfectly with the classical adaptive control scheme.

By assuming that the order of the plant (that is, the number of state variables) is known and all state variables are measurable, we have

$$\mathbf{x}(k+1) = \mathbf{f}(\mathbf{x}(k), u(k)) \tag{3.26}$$

where $\mathbf{x}(k+1)$ is the state at time k+1, $\mathbf{x}(k)$ is the state at time k, and u(k) is the control signal at time k (assuming for simplicity that u(k) is a scalar). Similarly, the state at time k+2 is expressed as

$$\mathbf{x}(k+2) = \mathbf{f}(\mathbf{x}(k+1), u(k+1)) = \mathbf{f}(\mathbf{f}(\mathbf{x}(k), u(k)), u(k+1)) \tag{3.27}$$

In general, we have

$$\mathbf{x}(k+n) = \mathbf{F}(\mathbf{x}(k), \mathbf{U}) \tag{3.28}$$

where n is the order of the plant, **F** is a multiple composite function of **f**, and **U** is the control actions from k to k+n-1, which is equal to

$$[u(k), u(k+1), ..., u(k+n-1)]^T$$

The preceding equation points out the fact that given the control input u from time k to k+n-1, the state of the plant will move from $\mathbf{x}(k)$ to $\mathbf{x}(k+n)$ in exactly n time steps. Furthermore, we assume that the inverse dynamics of the plant do exist, that is, **U** can be expressed as an explicit function of $\mathbf{x}(k)$ and $\mathbf{x}(k+n)$:

$$\mathbf{U} = \mathbf{G}(\mathbf{x}(k), \mathbf{x}(k+n)) \qquad (3.29)$$

This equation essentially says that there exists a unique input sequence **U**, specified by mapping **G**, that can drive the plant from state $\mathbf{x}(k)$ to $\mathbf{x}(k+n)$ in n time steps. The problem now becomes how to find the inverse mapping **G**.

Although the inverse mapping **G** in Equation (3.29) exists by assumption, it does not always have an analytically closed form. Therefore, instead of looking for methods of solving Equation (3.29) explicitly, we can use an adaptive network or ANFIS with 2n inputs and n outputs to approximate the inverse mapping **G** according to the generic training data pairs

$$[\mathbf{x}(k)^T, \mathbf{x}(k+n)^T ; \mathbf{U}^T] \qquad (3.30)$$

Figure 3.8 illustrates the situation in which n is equal to 1. Figure 3.8 (a) shows a plant block in which the plant output x(k+1) is a function of a previous state x(k) and input u(k); we use z^{-1} block to represent the unit-time delay operator. Figure 3.8 (b) is the block diagram during the training phase; Figure 3.8 (c) is the block diagram during the application phase.

Assume that the adaptive network truly imitates the input-output mapping of the inverse dynamics **G**. Then, given the current state $\mathbf{x}(k)$ and the desired future state $\mathbf{x}_d(k+n)$, the adaptive network will generate an estimated $\hat{\mathbf{U}}$:

$$\hat{\mathbf{U}} = \hat{\mathbf{G}}(\mathbf{x}(k), \mathbf{x}_d(k+n)) \qquad (3.31)$$

After n steps, this control sequence can bring the state $\mathbf{x}(k)$ to the desired state $\mathbf{x}_d(k+n)$, assuming that the adaptive network function $\hat{\mathbf{G}}$ is exactly the same as the inverse mapping **G**. This application phase is shown in the block diagram of Figure 3.8 (b). If the future desired state $\mathbf{x}_d(k+n)$ is not available in advance, we can use the current desired state $\mathbf{x}_d(k)$ in Figure 3.8 (b). This implies that the current desired state will appear after n time steps and the whole system behaves like a pure n-step time delay system.

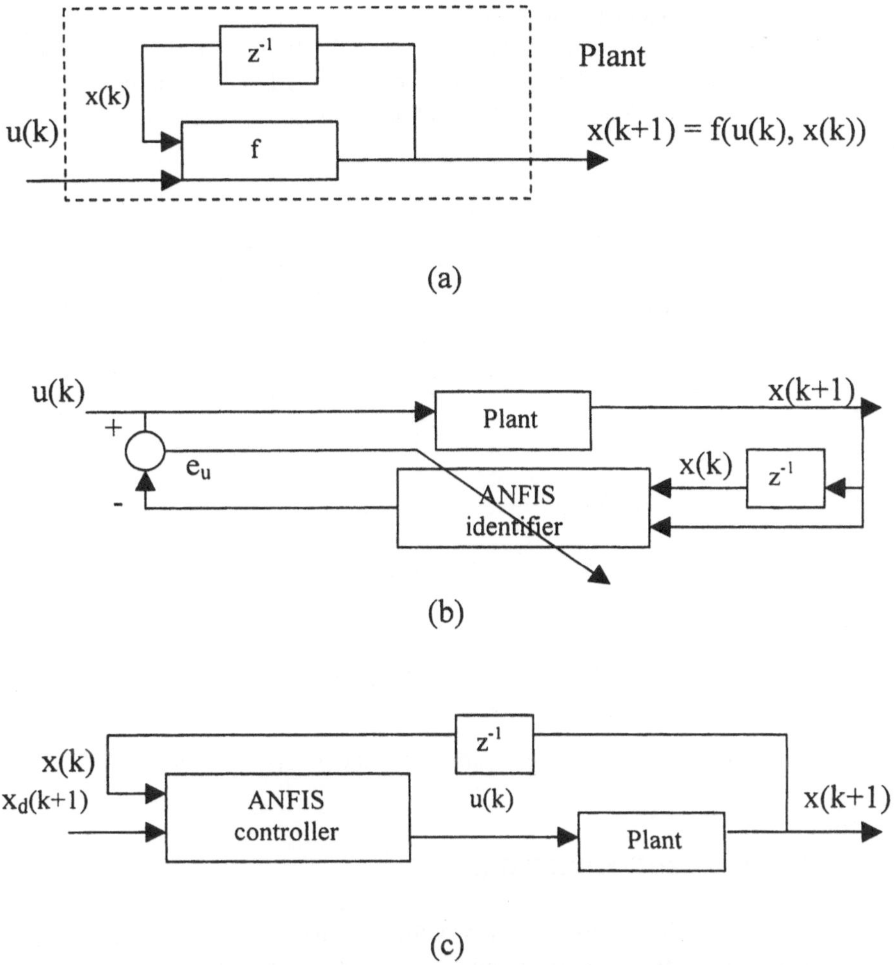

Figure 3.8 Block diagram for the inverse learning method: (a) plant block; (b) training phase; (c) application phase

When $\widehat{\mathbf{G}}$ is not close to **G**, the control sequence $\widehat{\mathbf{U}}$ cannot bring the state to $\mathbf{x}_d(k+n)$ in exactly the next n time step. As more data pairs are used to refine the parameters in the adaptive network, $\widehat{\mathbf{G}}$ will become closer to **G** and the control will be more and more accurate as the training process goes on.

For off-line applications, we have to collect a set of training data pairs and then train the adaptive network in the batch mode. For on-line applications to deal with time-varying systems, the control actions in Equation (3.31) are generated every n time steps while on-line learning occurs at every time step. Alternatively, we can generate the control sequence at every time step and apply only the first component to the plant. Figure 3.9 is a block diagram for on-line learning when n is equal to 1. The dashed line in the figure indicates that the two ANFIS blocks are exact duplicates of each other. (For simplicity, we have removed the unit-time delay operator from this figure).

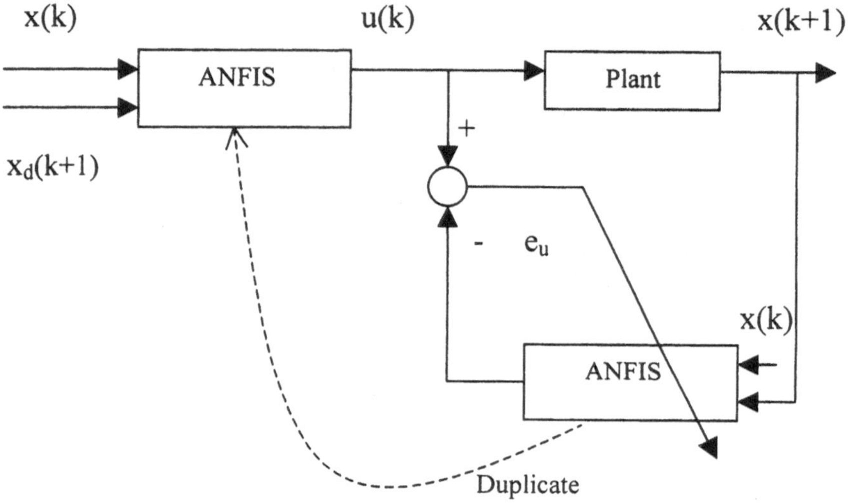

Figure 3.9 Block diagram for on-line inverse learning

3.3.2 Specialized Learning

A major problem with inverse learning is that an inverse model does not always exist for a given plant. Moreover, inverse learning is an indirect approach that tries to minimize the network output error instead of the overall system error (defined as the difference between desired and actual trajectories). "Specialized learning" (Psaltis, Sideris & Yamamura, 1988) is an alternative method that tries to minimize the system error directly by backpropagating error signals through the plant block. The price that we pay is that we need to know more about the plant under consideration.

Figure 3.10 illustrates the most basic type of specialized learning, Figure 3.10 (a) is the plant block (assuming its order is 1), and Figure 3.10 (b) indicates the training of the ANFIS controller. The ANFIS parameters are updated to reduce the system error $e_x(k)$, which is defined as the difference between the system's output $\mathbf{x}(k)$ and the desired output $\mathbf{x}_d(k)$.

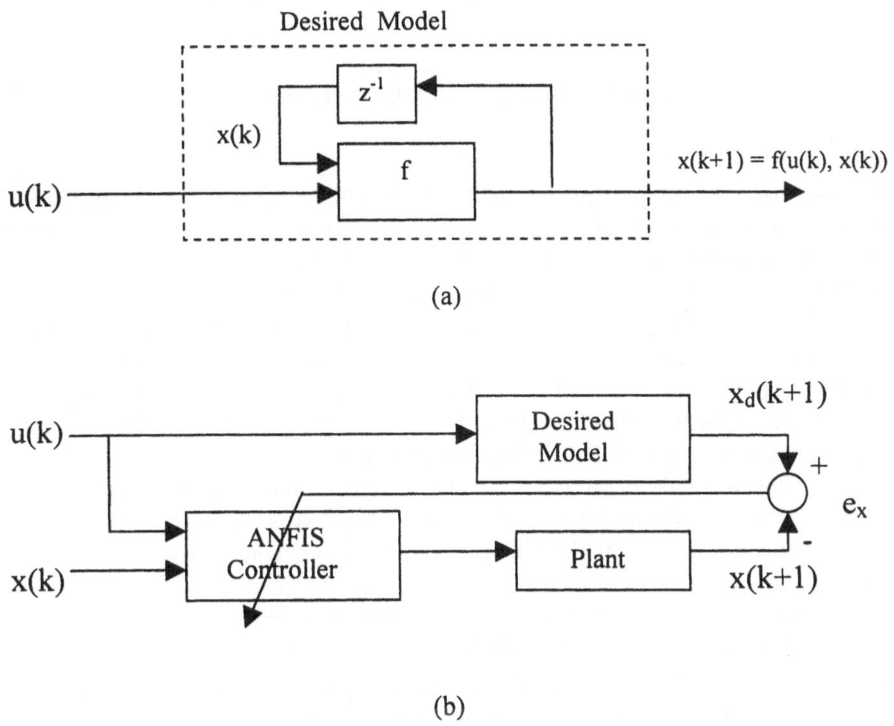

Figure 3.10 (a) Desired model block; (b) specialized learning with model referencing

To be more specific, let the plant dynamics be specified by

$$\mathbf{x}(k+1) = \mathbf{f}(\mathbf{x}(k), v(k))$$

and the ANFIS output be denoted as

$$\hat{v}(k) = F(\mathbf{x}(k), u(k), \theta) \tag{3.32}$$

where θ is a parameter vector to be updated. If we set the ANFIS output as the plant's input, then $v(k) = \hat{v}(k)$ and we have a closed-loop system specified by

$$\mathbf{x}(k+1) = \bar{\mathbf{f}}(\mathbf{x}(k), F(\mathbf{x}(k), u(k), \theta))$$

The objective of specialized learning is to minimize the difference between the closed-loop system and the desired model. Hence we can define an error measure:

$$J(\theta) = \sum_k \|\mathbf{f}(\mathbf{x}(k), F(\mathbf{x}(k), u(k), \theta)) - \mathbf{x}_d(k+1)\|^2 \qquad (3.33)$$

We can use backpropagation or steepest descent to update θ to minimize the above error measure. To find the derivative of $J(\theta)$ with respect to θ, we need to know the derivative of **f** with respect to its second argument. In other words, to backpropagate error signals through the plant block in Figure 3.10 (b), we need to know the "Jacobian matrix" of the plant, where the element at row i and column j is equal to the derivative of the plant's ith output with respect to its jth input. This usually implies that we need a model for the plant and the Jacobian matrix obtained from the model, which could be a neural network, an ANFIS, or another appropriate mathematical description of the plant.

For a single-input plant, if the Jacobian matrix is not easily found directly, a crude estimate can be obtained by approximating it directly from the changes in the plant's input and output(s) during two consecutive time instants. Other methods that aim at using an approximate Jacobian matrix to achieve the same learning effects can be found in Chen and Pao (1989).

It is not always convenient to specify the desired plant output $\mathbf{x}_d(k)$ at every time instant k. As a standard approach in model reference adaptive Control, the desired behavior of the overall system can be implicitly specified by a model that is able to achieve the control goal satisfactorily. Let the desired model be specified by

$$\mathbf{x}_d(k+1) = \bar{\mathbf{f}}(\mathbf{x}(k), u(k))$$

Then the error measure in Equation (3.33) becomes

$$J(\theta) = \sum_k \|\mathbf{f}(\mathbf{x}(k), F(\mathbf{x}(k), u(k), \theta)) - \bar{\mathbf{f}}(\mathbf{x}(k), u(k))\|^2 \qquad (3.34)$$

Again, we still need the Jacobian matrix of the plant to do backpropagation.

Note that the ANFIS controller in Equation (3.32) represents the most general situation. More commonly, the ANFIS controller is a function of $\mathbf{x}(k)$ and θ only and the input to the plant v(k) is expressed as the difference between the command signal u(k) and ANFIS output, as follows:

$$\hat{v}(k) = u(k) - F(\mathbf{x}(k), \theta)\ .$$

3.4 Adaptive Model-Based Neuro-Control

This section briefly reviews various approaches in current adaptive neuro-control design (Odmivar & Elliot, 1997). Although there are other ways to classify these approaches (e.g., Hunt, Sbarbaro, Zbikowski & Gawthrop, 1992) this section nevertheless adopts one similar to adaptive control theory: 1) indirect neuro-control and 2) direct neuro-control.

In the indirect neuro-control scheme, a neural network does not send a control signal "directly" to the process. Instead, a neural network is often used as an indirect process characteristics indicator. This indicator can be a process model that mimics the process behavior or a controller auto-tuner that produces appropriate controller settings based upon the process behavior. In this category, the neuro-control approaches can be roughly distinguished as follows: 1) neural network model-based control, 2) neural network inverse model-based control, and 3) neural network auto-tuner development.

In the direct neuro-control scheme, a neural network is employed as a feedback controller, and it sends control signals "directly" to the process. Depending on the design concept, the direct neuro-control approaches can be categorized into: 1) controller modelling, 2) model-free neuro-control design, 3) model-based neuro-control design, and 4) robust model-based neuro-control design.

Regardless of these distinctions, a unifying framework for neuro-control is to view neural network training as a non-linear optimization problem,

$$\text{NN:} \min_{w} J(w) \tag{3.35}$$

in which one tries to find an optimal representation of the neural network that minimizes an objective function J over the network weight space w. Here, NN indicates that the optimization problem formulation involves a neural network. The role a neural network plays in the objective function is then a key to distinguishing the various neuro-control design approaches.

3.4.1 Indirect Neuro-Control

The most popular control system application of neural networks is to use a neural network as an input-output process model. This approach is a data-driven supervised learning approach, i.e., the neural network attempts to mimic an existing process from being exposed to the process data (see Figure 3.11). The most commonly adopted model structure for such a purpose is the non-linear auto-regressive and moving average with exogenous inputs (known as NARMAX) model or a simpler NARX (Su, McAvoy & Werbos, 1992). Alternatively, one can choose to identify a continuous-time model with a dynamic neural network. Regardless of the model structure and the control strategy, the neuro-control design in this case can be conceptually stated as follows:

$$\text{NN: } \min_{w} F\{\, y_p - y_n(w, \ldots) \,\} \tag{3.36}$$

where y_p stands for plant/process output, y_n for neural network output, and w for neural network weights. Here F is a functional that measures the performance of the optimization process. It is usually an integral or sum of the prediction errors between y_p and y_n. For example, in this model development stage, process inputs and outputs $\{u_p, y_p\}$ are collected over a finite period of time and used for neural network training.

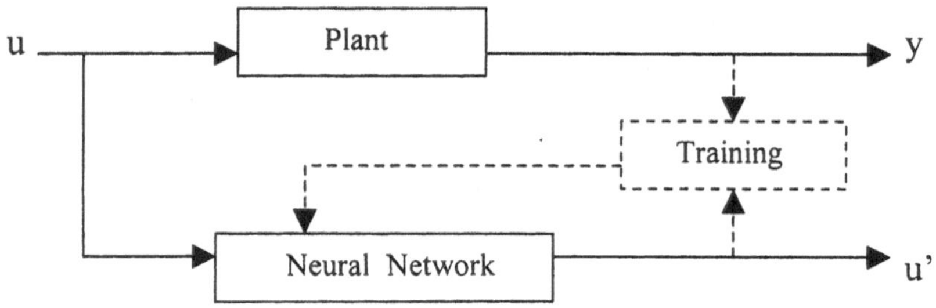

Figure 3.11 Neural Network as a black-box model of a process

At the implementation stage, nevertheless, the neural network model cannot be used alone. It must be incorporated with a model-based control scheme. In the chemical process industry, for example, a neural network is usually employed in a non-linear model predictive control (MPC) scheme (Su & McAvoy, 1993). Figure 3.12 illustrates the block diagram of an MPC control system. In fact, the MPC control is also an optimization problem.

The optimization problem here can be expressed as follows:

$$\min_{u} F' \{\, y^* - y_n (u, \ldots) \,\} \tag{3.37}$$

where y* designates the desired close-loop process output, u the process/model input or control signal, and y_n the predicted process output (by the neural network model). Here F' stands for an objective function that evaluates the closed-loop performance. For example, the optimization problem in the implementation stage is usually as follows:

$$\min_{u} \sum_{t} \| y^*(t) - y_n(t) - d(t) \|^2 \ , \quad y_n = \aleph\,(u, \ldots) \qquad (3.38)$$

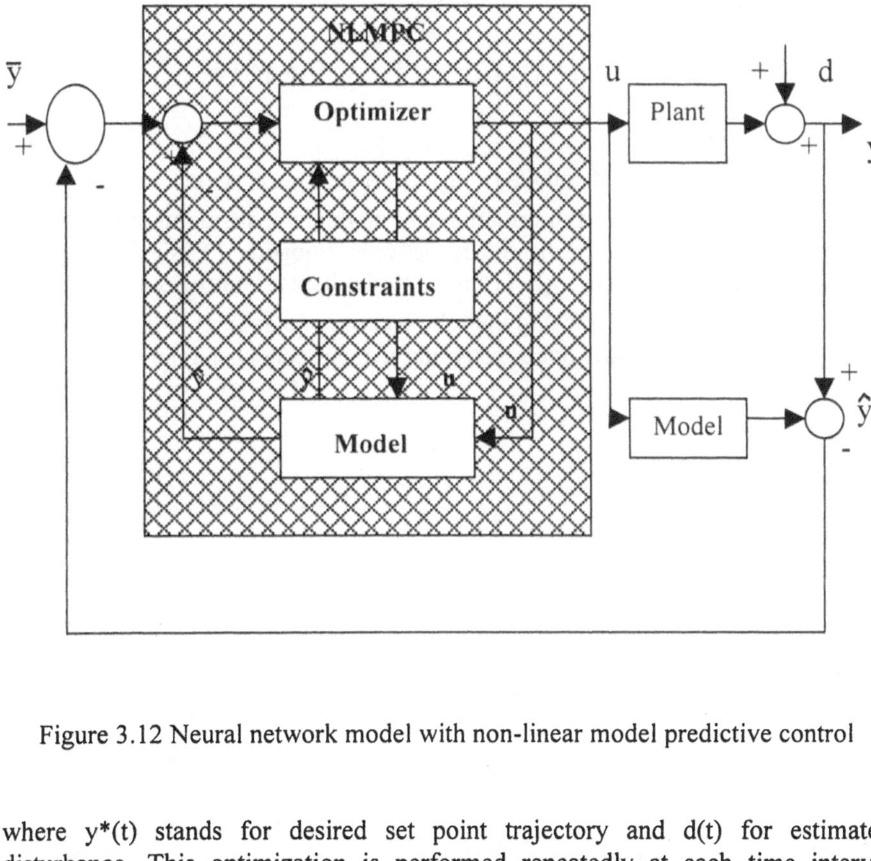

Figure 3.12 Neural network model with non-linear model predictive control

where y*(t) stands for desired set point trajectory and d(t) for estimated disturbance. This optimization is performed repeatedly at each time interval during the course of feedback control. Although the constrains are not particularly of interest in the discussion, one advantage of this indirect control design approach over the direct ones is that the constraints can be incorporated when solving the above optimization problem.

In some cases, a certain degree of knowledge, about the process might be available, such as model structure or particular physical phenomena that are well understood. In this case, a full black-box model might not be most desirable. For example, if the structure of the process model is available, values for the associated parameters can be determined by a neural network. Examples of these parameters can be time constants, gains, and delays or physical parameters such as diffusion rates and heat transfer coefficients. When model structure is not known a priori, neural networks can be trained to select elements of a model structure from a predetermined set. Lastly, in other cases where model structure is partially known, neural networks can also be integrated with such a partial model so that the process can be better model (see Figure 3.13).

For illustration purposes, the parametric or partial neural network modelling problem can be formulated as follows:

$$\text{NN:} \min_{w} F\{ y_p - y_m(\theta, \ldots) \} \quad , \quad \theta = \aleph(w, \ldots) \qquad (3.39)$$

where y_m is the predicted output from the model and θ stands for the process parameters, model structural information and other elements required to complete the model. Notice the only difference between Equation (3.39) and Equation (3.36) is that y_m replaces y_n. From a model-based control standpoint, this approach is essentially identical to the full black-box neural network model except that the neural network does not directly mimic the process behavior.

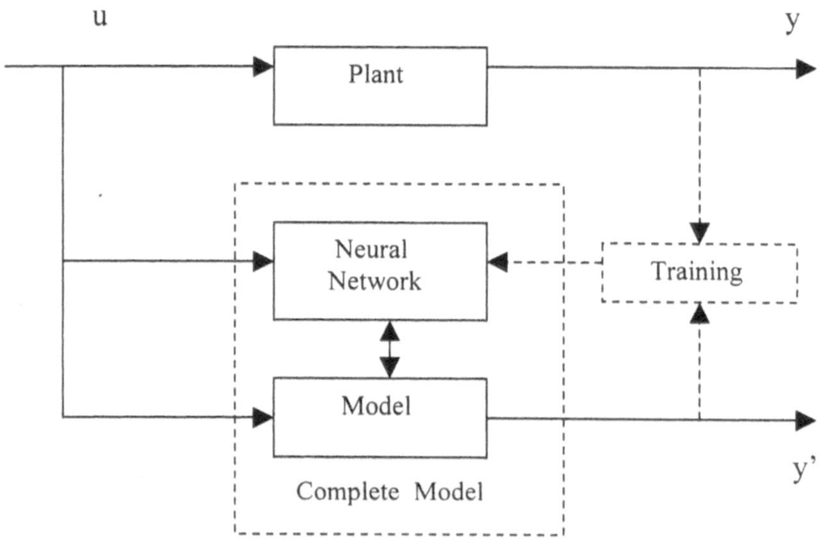

Figure 3.13 A neural network can be a parameter estimator, model structure selector, or a partial element of a physical model

A neural network can be trained to develop an inverse model of the plant. The network input is the process output, and the network output is the corresponding process input (see Figure 3.14). In general, the optimization problem can be formulated as

$$\text{NN:} \min_{w} F \{ u^*_{p-1} - u_n (w, \dots)\} \tag{3.40}$$

where u^*_{p-1} is the process inputs. Typically, the inverse model is a steady state/static model, which can be used for feedforward control. Given a desired process set point y*, the appropriate steady-state control signal u* for this set point can be immediately known:

$$u^* = \aleph (y^*, \dots) \tag{3.41}$$

Successful applications of inverse modelling are discussed in (Miller, Sutton & Werbos, 1995). Obviously, an inverse model exists only when the process behaves monotonically as a "forward" function at steady state.

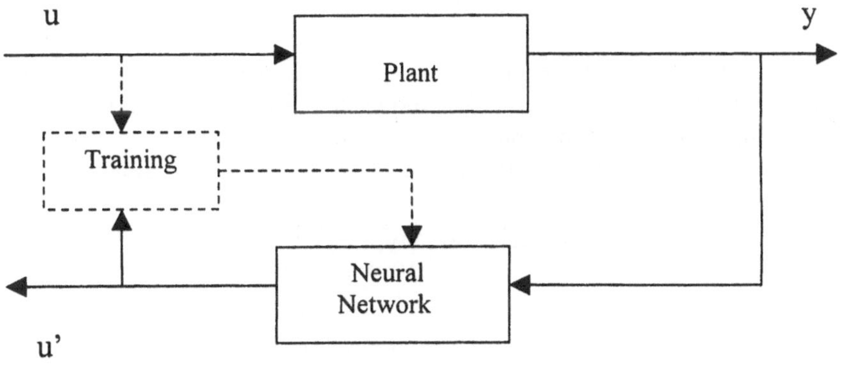

Figure 3.14 A neural network inverse model

As in the previous case where neural networks can be used to estimate parameters of a known model, they can also be used to estimate tuning parameters of a controller whose structure is known a priori. A controller's tuning parameter estimator is often referred to as an autotuner. The optimization problem in this case can be formulated as follows:

$$\text{NN: } \min_{w} F\ \{\ \eta^* - \eta_n\ (w, \ldots\ )\} \qquad (3.42)$$

where η^* denotes the controller parameters as targets and η_n stands for the predicted values by the neural network. Network input can comprise sampled process data or features extracted from it. However, these parameters η cannot be uniquely determined from the process characteristics. They also depend on the desired closed-loop control system characteristics. Usually, the controller parameters are solutions to the following closed-loop control optimization:

$$\min_{\eta} F'\ \{\ y^* - y_{p/m}\ (u, \ldots\ )\} \quad ; \quad u = C(\eta, \ldots\ ) \qquad (3.43)$$

where C is a controller with a known structure. Here, $y_{p/m}$ denotes that either a process or a model can be employed in this closed-loop control in order to find the target controller C.

3.4.2 Direct Neuro-Control

Among the four direct neuro-control schemes, the simplest for neuro-controller development is to use a neural network to model an existing controller (see Figure 3.15). The input to the existing controller is the training input to the network and the controller output serves as the target. This neuro-control design can be formulated as follows:

$$\text{NN: } \min_{w} F\ \{u^*_c - u_n\ (w, \ldots\ )\} \qquad (3.44)$$

where u^*_c is the output of an existing controller C*. Usually, the existing controller C* can be a human operator or it can be obtained via

$$\min_{c} F'\ \{y^* - y_{p/m}\ (u, \ldots\ )\} \quad ; \quad u = C(\ \ldots\ ) \qquad (3.45)$$

Like a process model, a controller is generally a dynamical system and often comprises integrators or differentiators. If a feedforward network is used to model the existing controller, dynamical information must be explicitly provided as input to the network.

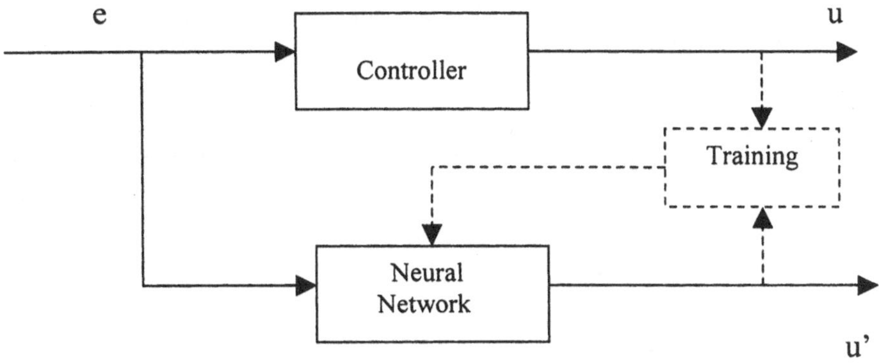

Figure 3.15 The simplest approach to neuro-control is to use a neural network to model an existing controller

While the benefits of this approach may be apparent when the existing controller is a human, its utility may be limited. It is applicable only when an existing controller is available, which is the case in many applications. Staib & Staib (1992) discuss how it can be effective in a multistage training process.

In the absence of an existing controller, some researchers have been inspired by the way a human operator learns to "control/operate" a process with little or no detailed knowledge of the process dynamics. Thus they have attempted to design controllers that by adaptation and learning can solve difficult control problems in the absence of process models and human design effort. In general, this model-free neuro-control can be stated as:

$$\text{NN:} \min_{w} F\{y^* - y_p(u, \dots)\} \quad , \quad u = \aleph(w, \dots) \qquad (3.46)$$

where y_p is the output from the plant. The key feature of this direct adaptation control approach is that a process model is neither known in advance nor explicitly developed during control design. Figure 3.16 is a typical representation of this class of control design.

The first work in this area was the "adaptive critic" algorithm proposed by Barto et al. (1983). Such an algorithm can be seen as an approximate version of dynamic programming. In this work, they posed a well-known cart-pole balancing problem and demonstrated their design concept. In this class of control design, limited/poor information is often adopted as an indication of performance criteria.

For example, the objective is the cart-pole balancing problem is simply to maintain the pole in a near-upright balanced position for as long as possible. The instructional feedback is limited to a "failure" signal when the controller fails to hold the pole in an upright position. The cart-pole problem has become a popular test-bed for explorations of the model-free control design concept.

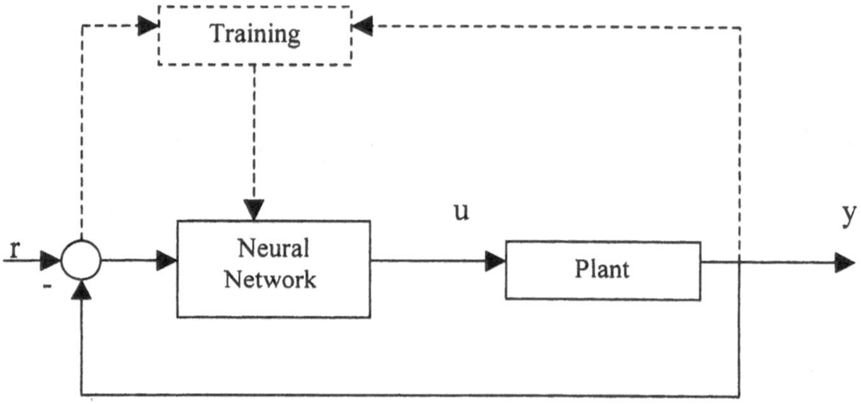

Figure 3.16 The model-free control design concept

Despite its historical importance and intuitive appeal, model-free adaptive neuro-control is not appropriate for most real-world applications. The plant is most likely out of control during the learning process, and few industrial processes can tolerate the large number of "failures" needed to adapt the controller.

From a practical perspective, one would prefer to let failures take place in a simulated environment (with a model) rather than in a real plant even if the failures are not disastrous or do not cause substantial losses. As opposed to the previous case, this class of neuro-control design is referred to as "model-based neuro-control design". Similar to Equation (3.46), as a result, the problem formulation becomes

$$\text{NN: } \min_{w} F\,\{y^* - y_m(u, \ldots)\}\quad,\quad u = \aleph\,(w, \ldots) \qquad (3.47)$$

Here, y_p in Equation (3.46) is replaced by y_m-the model's output. In this case, knowledge about the processes of interest is required. As can be seen in Figure 3.17, a model replaces the plant/process in the control system.

If a process model is not available, one can first train a second neural network to model the plant dynamics. In the course of modelling the plant, the plant must be operated "normally" instead of being driven out of control. After the modelling stage, the model can then be used for control design. If a plant model is already available, a neural network controller can then be developed in a simulation in which failures cannot cause any loss but that of computer time. A neural network controller after extensive training in the simulation can then be installed in the actual control system.

In fact, these "model-based neuro-control design" approaches have not only proven effective in several studies (Troudet, 1991), but also have already produced notable economic benefits (Staib, 1993). Nevertheless, the quality of control achieved with this approach depends crucially on the quality of the process model. If a model is not accurate enough, the trained neuro-controller is unlikely to perform satisfactorily on the real process. Without an on-line adaptive component, this neuro-controller does not allow for plant drifts or other factors that could adversely affect the performance of the control system.

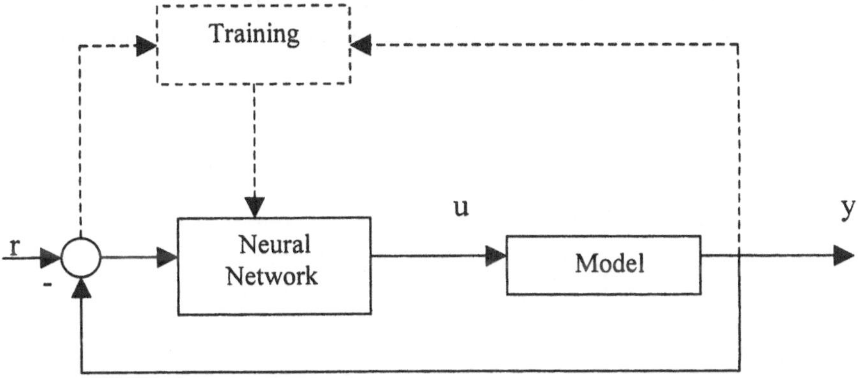

Figure 3.17 A model replaces the plant/process in the control system during the control design phase

The neuro-controller approaches discussed above still share a common shortcoming: A neural network must be trained for every new application. Network retraining is needed even with small changes in the control criterion, such as changes in the relative weighting of control energy and tracking response, or if the controller is to be applied to a different but similar processes. In order to avoid such drawbacks, the concept of "robustness" is naturally brought into the design of a neuro-controller. In robust model-based neuro-control design, a family of process models is considered instead of just a nominal one (see Figure 3.18).

Often such a family is specified by a range of noise models or range of the process parameters. Robust neuro-control design can be formulated as follows:

$$\text{NN: } \min_{w} F \{y^* - y_{mi}(u, \ldots)\} \;,\; u = \aleph(w, \ldots), \quad \forall\, m_i \in \mathbf{M} \quad (3.48)$$

where m_i stands for the ith member of the model family **M**. Ideally, the real process to be controlled should belong to this family as well so that the controller is robust not only for the model but also for the real process.

Two aspects of robustness are commonly distinguished. Robust Stability refers to a control system that is stable (qualitatively) over the entire family of processes, whereas robust performance refers to (quantitative) performance criteria being satisfied over the family (Morari & Zafiriou, 1989). Not surprisingly, there is a tradeoff to achieve robustness. By optimizing a neural network controller based upon a fixed (and accurate) process model, high performance can be achieved as long as the process remains invariant, but at the likely cost of brittleness. A robust design procedure, on the other hand, is not likely to achieve the same level of nominal performance but will be less sensitive to process drifts, disturbances, and other sources of process-model mismatch.

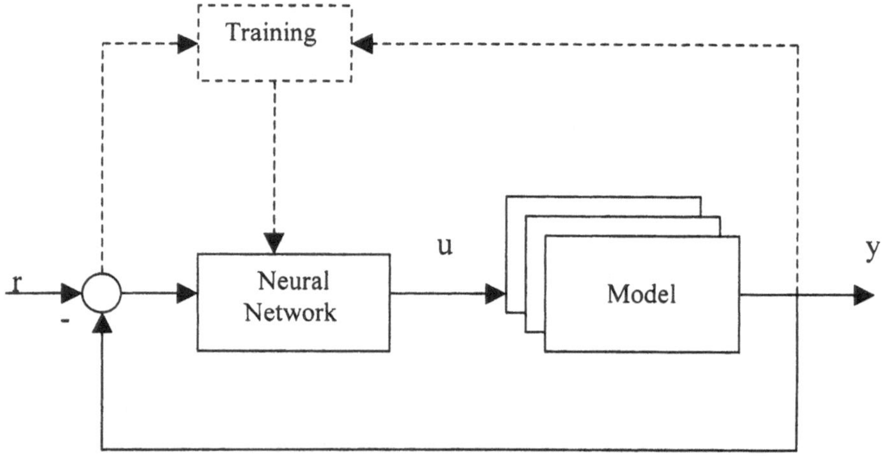

Figure 3.18 Robust model-based neuro-control

3.4.3 Parameterized Neuro-Control

All the above neuro-control approaches share a common shortcoming-the need for extensive application-specific development efforts. Each application requires the optimization of the neural network controller and may also require process model identification. The expense in time and computation has been a significant barrier to widespread implementation of neuro-control systems.

In an attempt to avoid application-specific development, a new neuro-control design concept-parameterized neuro-control (PNC)-has evolved (Samad & Foslien, 1994). Figure 3.19 illustrates this PNC strategy. The PNC controller is equipped with parameters that specify process characteristics and those that provide performance criterion information. For illustration purposes, a PNC can be conceptually formulated as follows:

$$\begin{aligned} &\text{NN: } \min_{w} F(\varepsilon)\,\{y^* - y_{mi}(\theta, u, \ldots)\}\ , \\ &u = \aleph\,(w, \theta, \varepsilon, \ldots)\ ,\ \forall\, m_i(\theta) \in \mathbf{M}(\theta) \end{aligned} \tag{3.49}$$

where ε designates the parameter set that defines the space of performance criteria, θ stands for the process parameter set, θ is the estimates for process parameters, and again $\mathbf{M}(\theta)$ is a family of parameterized models $m_i(\theta)$ in order to account for errors in process parameters estimates θ.

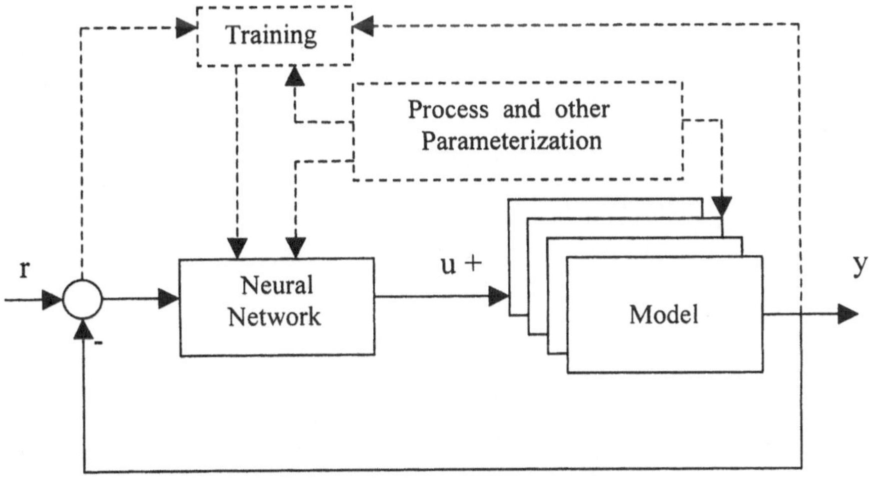

Figure 3.19 Parameterized Neuro-Control

In fact, the two additional types of parameters (ε and θ) make a PNC generic. A PNC is generic in two respects: 1) the process model parameters θ facilitate its application to different processes and 2) the performance parameters ε allow its performance characteristics to be adjustable, or "tunable". For example, if a PNC is designed for first-order plus delay processes, the process parameters (i.e., process gain, time constant, and dead time) will be adjustable parameters to this PNC. Once developed, this PNC requires no application-specific training or adaptation when applied to a first-order plus delay process. It only requires estimates of these process parameters. These estimates do not have to be accurate because the robustness against such inaccuracy is considered in the design phase.

3.5 Summary

In this chapter, we have presented the main ideas underlying Neural Networks and the application of this powerful computational theory to general control problems. We have discussed in some detail the backpropagation learning algorithm for feedforward networks, the integration of fuzzy logic techniques to neural networks to form powerful adaptive neuro-fuzzy inference systems and the basic concepts and current methods of neuro-fuzzy control. At the end, we also gave some remarks about adaptive neuro-control and model-based control of non-linear dynamical systems. In the following chapters, we will show how neural network techniques (in conjunction with other techniques) can be applied to solve real world complex problems of control. This chapter will serve as a basis for the new hybrid intelligent control methods that will be described in Chapter 7 of this book.

Chapter 4

Genetic Algorithms and Simulated Annealing

This chapter introduces the basic concepts and notation of genetic algorithms and simulated annealing, which are two basic search methodologies that can be used for modelling and simulation of complex non-linear dynamical systems. Since both techniques can be considered as general purpose optimization methodologies, we can use them to find the mathematical model which minimizes the fitting errors for a specific problem. On the other hand, we can also use any of these techniques for simulation if we exploit their efficient search capabilities to find the appropriate parameter values for a specific mathematical model. We also describe in this chapter the application of genetic algorithms to the problem of finding the best neural network or fuzzy system for a particular problem. We can use a genetic algorithm to optimize the weights or the architecture of a neural network for a particular application. Alternatively, we can use a genetic algorithm to optimize the number of rules or the membership functions of a fuzzy system for a specific problem. These are two important application of genetic algorithms, which will be used in later chapters to design intelligent intelligent systems for controlling real world dynamical systems.

Genetic algorithms and simulated annealing have been used extensively for both continuous and discrete optimization problems (Jang, Sun & Mizutani, 1997). Common characteristics shared by these methods are described next.

- <u>Derivative freeness</u>: These methods do not need functional derivative information to search for a set of parameters that minimize (or maximize) a given objective function. Instead they rely exclusively on repeated evaluations of the objective function, and the subsequent search direction after each evaluation follows certain heuristic guidelines.

- <u>Heuristic guidelines</u>: The guidelines followed by these search procedures are usually based on simple intuitive concepts. Some of these concepts are motivated by so-called nature's wisdom, such as evolution and thermodynamics.

- Flexibility: Derivative freeness also relieves the requirement for differentiable objective functions, so we can use as complex an objective function as a specific application might need, without sacrificing too much in extra coding and computation time. In some cases, an objective function can even include the structure of a data-fitting model itself, which may be a fuzzy model.

- Randomness: These methods are stochastic, which means that they use random number generators in determining subsequent search directions. This element of randomness usually gives rise to the optimistic view that these methods are "global optimizers" that will find a global optimum given enough computing time. In theory, their random nature does make the probability of finding an optimal solution nonzero over a fixed amount of computation time. In practice, however, it might take a considerable amount of computation time.

- Analytic opacity: It is difficult to do analytic studies of these methods, in part because of their randomness and problem-specific nature. Therefore, most of our knowledge about them is based on empirical studies.

- Iterative nature: These techniques are iterative in nature and we need certain stopping criteria to determine when to terminate the optimization process. Let K denote an iteration count and f_k denote the best objective function obtained at count k; common stopping criteria for a maximization problem include the following:

    1) Computation time: a designated amount of computation time, or number of function evaluations and/or iteration counts is reached.
    2) Optimization goal: f_k is less than a certain preset goal value.
    3) Minimal improvement: $f_k - f_{k-1}$ is less than a preset value.
    4) Minimal relative improvement: $(f_k - f_{k-1})/ f_{k-1}$ is less than a preset value.

Both genetic algorithms (GAs) and simulated annealing (SA) have been receiving increasing amounts of attention due to their versatile optimization capabilities for both continuous and discrete optimization problems. Moreover, both of them are motivated by so-called "nature's wisdom": GAs are based on the concepts of natural selection and evolution; while SA originated in annealing processes found in thermodynamics and metallurgy.

4.1 Genetic Algorithms

Genetic algorithms (GAs) are derivative-free optimization methods based on the concepts of natural selection and evolutionary processes (Goldberg, 1989). They were first proposed and investigated by John Holland at the University of Michigan (Holland, 1975). As a general-purpose optimization tool, GAs are moving out of academia and finding significant applications in many areas. Their popularity can be attributed to their freedom from dependence on functional derivatives and their incorporation of the following characteristics:

- GAs are parallel-search procedures that can be implemented on parallel processing machines for massively speeding up their operations.
- GAs are applicable to both continuous and discrete (combinatorial) optimization problems.
- GAs are stochastic and less likely to get trapped in local minima, which inevitably are present in any optimization application.
- GAs' flexibility facilitates both structure and parameter identification in complex models such as fuzzy inference systems or neural networks.

GAs encode each point in a parameter (or solution) space into a binary bit string called a "chromosome", and each point is associated with a "fitness value" that, for maximization, is usually equal to the objective function evaluated at the point. Instead of a single point, GAs usually keep a set of points as a "population", which is then evolved repeatedly toward a better overall fitness value. In each generation, the GA constructs a new population using "genetic operators" such as crossover and mutation; members with higher fitness values are more likely to survive and to participate in mating (crossover) operations. After a number of generations, the population contains members with better fitness values; this is analogous to Darwinian models of evolution by random mutation and natural selection. GAs and their variants are sometimes referred to as methods of "population-based optimization" that improve performance by upgrading entire populations rather than individual members. Major components of GAs include encoding schemes, fitness evaluations, parent selection, crossover operators, and mutation operators; these are explained next.

<u>Encoding schemes</u>: These transform points in parameter space into bit string representations. For instance, a point (11, 4, 8) in a three-dimensional parameter space can be represented as a concatenated binary string:

$$\underbrace{1011}_{11}\ \underbrace{0100}_{4}\ \underbrace{1000}_{8}$$

in which each coordinate value is encoded as a "gene" composed of four binary bits using binary coding. other encoding schemes, such as gray coding, can also be used and, when necessary, arrangements can be made for encoding negative, floating-point, or discrete-valued numbers. Encoding schemes provide a way of translating problem-specific knowledge directly into the GA framework, and thus play a key role in determining GAs' performance. Moreover, genetic operators, such as crossover and mutation, can and should be designed along with the encoding scheme used for a specific application.

Fitness evaluation: The first step after creating a generation is to calculate the fitness value of each member in the population. For a maximization problem, the fitness value f_i of the ith member is usually the objective function evaluated at this member (or point). We usually need fitness values that are positive, so some kind of monotonical scaling and/or translation may by necessary if the objective function is not strictly positive. Another approach is to use the rankings of members in a population as their fitness values. The advantage of this is that the objective function does not need to be accurate, as long as it can provide the correct ranking information.

Selection: After evaluation, we have to create a new population from the current generation. The selection operation determines which parents participate in producing offspring for the next generation, and it is analogous to "survival of the fittest" in natural selection. Usually members are selected for mating with a selection probability proportional to their fitness values. The most common way to implement this is to set the selection probability equal to:

$$f_i / \sum_{k=1}^{k=n} f_k,$$

where n is the population size. The effect of this selection method is to allow members with above-average fitness values to reproduce and replace members with below-average fitness values.

Crossover: To exploit the potential of the current population, we use "crossover" operators to generate new chromosomes that we hope will retain good features from the previous generation. Crossover is usually applied to selected pairs of parents with a probability equal to a given "crossover rate". "One-point crossover" is the most basic crossover operator, where a crossover point on the genetic code is selected at random and two parent chromosomes are interchanged at this point. In "two-point crossover", two crossover points are selected and the part of the chromosome string between these two points is then swapped to generate two children. We can define n-point crossover similarly. In general, (n-1)-point crossover is a special case of n-point crossover. Examples of one-and two-point crossover are shown in Figure 4.1.

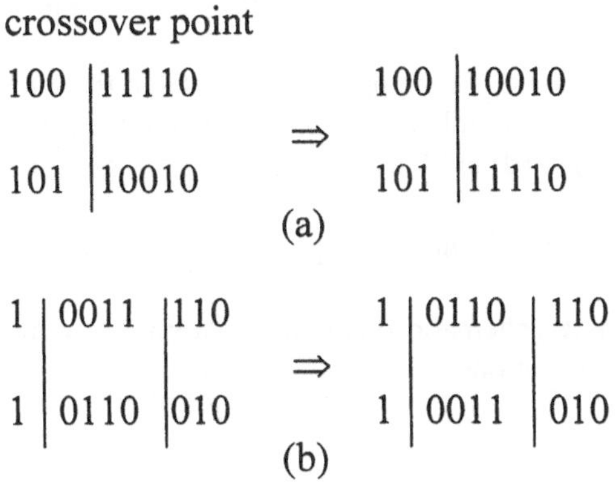

Figure 4.1 Crossover operators: (a) one-point crossover; (b) two-point crossover.

<u>Mutation</u>: Crossover exploits current gene potentials, but if the population does not contain all the encoded information needed to solve a particular problem, no amount of gene mixing can produce a satisfactory solution. For this reason, a "mutation" operator capable of spontaneously generating new chromosomes is included. The most common way of implementing mutation is to flip a bit with a probability equal to a very low given "mutation rate". A mutation operator can prevent any single bit from converging to a value throughout the entire population and, more important, it can prevent the population from converging and stagnating at any local optima. The mutation rate is usually kept low so good chromosomes obtained from crossover are not lost. If the mutation rate is high (above 0.1), GA performance will approach that of a primitive random search. Figure 4.2 provides an example of mutation.

Figure 4.2 Mutation operator.

In the natural evolutionary process, selection, crossover, and mutation all occur in the single act of generating offspring. Here we distinguish them clearly to facilitate implementation of and experimentation with GAs.

Based on the aforementioned concepts, a simple genetic algorithm for maximization problems is described next.

Step 1: Initialize a population with randomly generated individuals and evaluate the fitness value of each individual.

Step 2: Perform the following operations:

(a) Select two members from the population with probabilities proportional to their fitness values.

(b) Apply crossover with a probability equal to the crossover rate.

(c) Apply mutation with a probability equal to the mutation rate.

(d) Repeat (a) to (d) until enough members are generated to form the next generation.

Step 3: Repeat steps 2 and 3 until a stopping criterion is met.

Figure 4.3 is a schematic diagram illustrating how to produce the next generation from the current one.

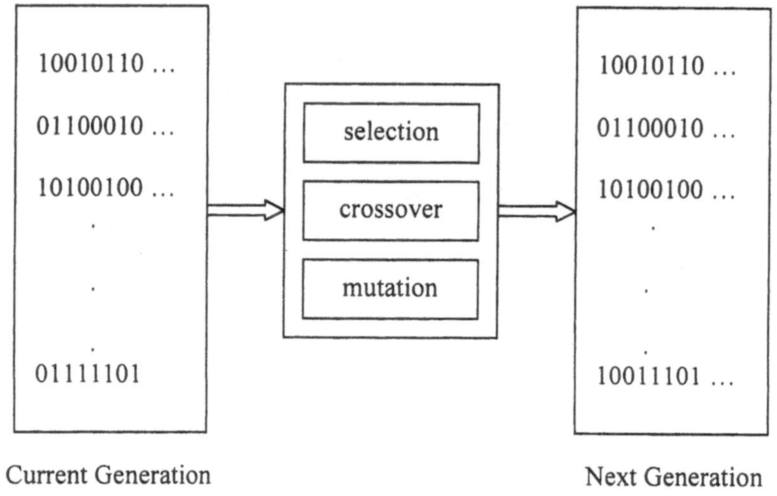

Figure 4.3 Producing the next generation in GAs.

4.2 Simulated Annealing

"Simulated Annealing" (SA) is another derivative-free optimization method that has recently drawn much attention for being as suitable for continuous as for discrete (combinational) optimization problems (Otten & Ginneken, 1989). When SA was first proposed (Kirkpatrick, Gelatt & Vecchi, 1983) it was mostly known for its effectiveness in finding near optimal solutions for large-scale combinatorial optimization problems, such as traveling salesperson problems and placement problems. Recent applications of SA and its variants (Ingber & Rosen, 1992) also demonstrate that this class of optimization approaches can be considered competitive with other approaches when there are continuous optimization problems to be solved.

Simulated annealing was derived from physical characteristics of spin glasses (Kirkpatrick, Gelatt & Vecchi, 1983). The principle behind simulated annealing is analogous to what happens when metals are cooled at a controlled rate. The slowly falling temperature allows the atoms in the molten metal to line themselves up and form a regular crystalline structure that has high density and low energy. But if the temperature goes down too quickly, the atoms do not have time to orient themselves into a regular structure and the result is a more amorphous material with higher energy.

In simulated annealing, the value of an objective function that we want to minimize is analogous to the energy in a thermodynamic system. At high temperatures, SA allows function evaluations at faraway points and it is likely to accept a new point with higher energy. This corresponds to the situation in which high-mobility atoms are trying to orient themselves with other nonlocal atoms and the energy state can occasionally go up. At low temperatures, SA evaluates the objective function only at local points and the likelihood of it accepting a new point with higher energy is much lower. This is analogous to the situation in which the low-mobility atoms can only orient themselves with local atoms and the energy state is not likely to go up again.

Obviously, the most important part of SA is the so-called "annealing schedule" or "cooling schedule", which specifies how rapidly the temperature is lowered from high to low values. This is usually application specific and requires some experimentation by trial-and-error.

Before giving a detailed description of SA, first we shall explain the fundamental terminology of SA.

Objective function: An objective function f(.) maps an input vector x into a scalar E:

$$E = f(x),$$

where each x is viewed as a point in an input space. The task of SA is to sample the input space effectively to find an x that minimizes E.

Generating function: A generating function g(. , .) specifies the probability density function of the difference between the current point and the next point to be visited. Specifically, Δx (= x_{new} - x) is a random variable with probability density function $g(\Delta x, T)$, where T is the temperature. For common SA (especially in combinatorial optimization applications), g(. , .) is usually a function independent of the temperature T.

Acceptance function: After a new point x_{new} has been evaluated, SA decides whether to accept or reject it based on the value of an acceptance function h(. , .).

The most frequently used acceptance function is the "Boltzmann probability distribution":

$$h(\Delta E, T) = \frac{1}{1 + \exp(\Delta E / (cT))} \qquad (4.1)$$

where c is a system-dependent constant, T is the temperature, and ΔE is the energy difference between x_{new} and x:

$$\Delta E = f(x_{new}) - f(x)$$

The common practice is to accept x_{new} with probability $h(\Delta E, T)$.

Annealing schedule: An annealing schedule regulates how rapidly the temperature T goes from high to low values, as a function of time or iteration counts. The exact interpretation of "high" and "low" and the specification of a good annealing schedule require certain problem-specific physical insights and/or trial-and-error. The easiest way of setting an annealing schedule is to decrease the temperature T by a certain percentage at each iteration.

The basic algorithm of simulated annealing is the following:

Step 1: Choose a start point x and set a high starting temperature T. Set the iteration count k to 1.

Step 2: Evaluate the objective function $E = f(x)$.

Step 3: Select Δx with probability determined by the generating function $g(\Delta x, T)$. Set the new point x_{new} equal to $x + \Delta x$.

Step 4: Calculate the new value of the objective function: $E_{new} = f(x_{new})$.

Step 5: Set x to x_{new} and E to E_{new} with probability determined by the acceptance function $h(\Delta E, T)$, where $\Delta E = E_{new} - E$.

Step 6: Reduce the temperature T according to the annealing schedule (usually by simply setting T equal to ηT, where η is a constant between 0 and 1).

Step 7: Increment iteration count k. If k reaches the maximum iteration count, stop the iterating. Otherwise, go back to step 3.

In conventional SA, also known as "Boltzmann machines", the generating function is a Gaussian probability density function:

$$g(\Delta x, T) = (2\pi T)^{-n/2} \exp[-\| \Delta x \|^2 / (2T)] \qquad (4.2)$$

where Δx ($= x_{new} - x$) is the deviation of the new point from the current one, T is the temperature, and n is the dimension of the space under exploration. It has been proven (Geman & Geman, 1984) that a Boltzman machine using the aforementioned generating function $g(\,.\,,\,.\,)$ can find a global optimum of f(x) if the temperature T is reduced no faster than $T_o / \ln k$.

Variants of Boltzmann machines include the "Cauchy machine" or "fast simulated annealing" (Szu & Hartley, 1987), where the generating function is the Cauchy distribution:

$$g(\Delta x) = \frac{T}{(\|\Delta x\|^2 + T^2)^{(n+1)/2}} \tag{4.3}$$

The fatter tail of the Cauchy distribution allows it to explore farther from the current point during the search process.

Another variant of the original SA, the so-called "very fast" simulated annealing (Ingber & Rosen, 1992), was designed for optimization problems in a constrained search space. Very fast simulated annealing has been reported to be faster than genetic algorithms on several test problems by the same authors.

4.3 Applications of Genetic Algorithms

The simple version of the genetic algorithm described in the previous section is very simple, but variations of this algorithm have been used in a large number of scientific and engineering problems and models (Mitchell, 1996). Some examples follow.

- Optimization: genetic algorithms have been used in a wide variety of optimization tasks, including numerical optimization and such combinatorial optimization problems as circuit layout and job-shop scheduling.
- Automatic Programming: genetic algorithms have been used to evolve computer programs for specific tasks, and to design other computational structures such as cellular automata and sorting networks.
- Machine Learning: genetic algorithms have been used for many machine learning applications, including classification and prediction tasks, such as the prediction of weather or protein structure. Genetic algorithms have also been used to evolve aspects of particular machine learning systems, such as weights for neural networks, rules for learning classifier systems or symbolic production systems, and sensors for robots.
- Economics: genetic algorithms have been used to model processes of innovation, the development of bidding strategies, and the emergence of economic markets.
- Immune Systems: genetic algorithms have been used to model various aspects of natural immune systems, including somatic mutation during an individual's lifetime and the discovery of multigene families during evolutionary time.
- Ecology: genetic algorithms have used to model ecological phenomena such as biological arms races, host-parasite coevolution, symbiosis, and resource flow.

- Social Systems: genetic algorithms have been used to study evolutionary aspects of social systems, such as the evolution of social behavior in insect colonies, and, more generally, the evolution of cooperation and communication in multi-agent systems.

This list is by no means exhaustive, but it gives the flavor of the kinds of things genetic algorithms have been used for, both in problem solving and in scientific contexts. Because of their success in these and other areas, interest in genetic algorithms has been growing rapidly in the last several years among researchers in many disciplines.

We will describe bellow the application of genetic algorithms to the problem of evolving neural networks, which is a very important problem in designing the particular neural network for a problem.

4.3.1 Evolving Neural Networks

Neural Networks are biologically motivated approaches to machine learning, inspired by ideas from neuroscience. Recently, some efforts have been made to use genetic algorithms to evolve aspects of neural networks (Mitchell, 1996).

In its simplest feedforward form, a neural network is a collection of connected neurons in which the connections are weighted, usually with real-valued weights. The network is presented with an activation pattern on its input units, such as a set of numbers representing features of an image to be classified. Activation spreads in a forward direction from the input units through one or more layers of middle units to the output units over the weighted connections. This process is meant to roughly mimic the way activation spreads through networks of neurons in the brain. In a feedforward network, activation spreads only in a forward direction, from the input layer through the hidden layers to the output layer. Many people have also experimented with "recurrent" networks, in which there are feedback connections between layers.

In most applications, the neural network learns a correct mapping between input and output patterns via a learning algorithm. Typically the weights are initially set to small random values. Then a set of training inputs is presented sequentially to the network. In the backpropagation learning procedure, after each input has propagated through the network and an output has been produced, a "teacher" compares the activation value at each output unit with the correct values, and the weights in the network are adjusted in order to reduce the difference between the network's output and the correct output. This type of procedure is known as "supervised learning", since a teacher supervises the learning by providing correct output values to guide the learning process.

There are many ways to apply genetic algorithms to neural networks. Some aspects that can be evolved are the weights in a fixed network, the network architecture (i.e., the number of neurons and their interconnections can change), and the learning rule used by the network.

4.3.1.1 Evolving Weights in a Fixed Network

David Montana and Lawrence Davis (1989) took the first approach of evolving the weights in a fixed network. That is, Montana and Davis were using the genetic algorithm instead of backpropagation as a way of finding a good set of weights for a fixed set of connections. Several problems associated with the backpropagation algorithm (e.g., the tendency to get stuck at local minima, or the unavailability of a "teacher" to supervise learning in some tasks) often make it desirable to find alternative weight training schemes.

Montana and Davis were interested in using neural networks to classify underwater sonic "lofargrams" (similar to spectrograms) into two classes: "interesting" and "not interesting". The networks were to be trained from a database containing lofargrams and classifications made by experts as to whether or not a given lofargram is "interesting". Each network had four input units, representing four parameters used by an expert system that performed the same classification. Each network had one output unit and two layers of hidden units (the first with seven units and the second with ten units). The networks were fully connected feedforward networks. In total there were 108 weighted connections between units. In addition, there were 18 weighted connections between the non-input units and a "threshold unit" whose outgoing links implemented the thresholding for each of the non-input units, for a total of 126 weights to evolve.

The genetic algorithm was used as follows. Each chromosome was a list of 126 weights. Figure 4.4 shows (for a much smaller network) how the encoding was done: the weights were read off the network in a fixed order (from left to right and from top to bottom) and placed in a list. Notice that each "gene" in the chromosome is a real number rather than a bit. To calculate the fitness of a given chromosome, the weights in the chromosome were assigned to the links in the corresponding network, the network was run on the training set (here 236 examples from the database), and the sum of the squares of the errors was returned. Here, an "error" was the difference between the desired output value and the actual output value. Low error meant high fitness in this case.

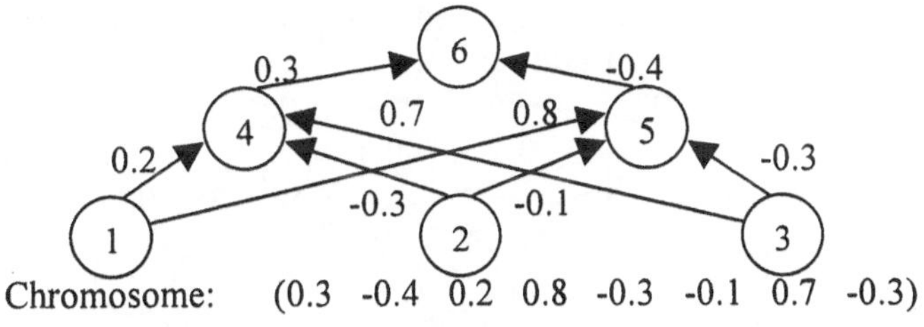

Figure 4.4 Encoding of network weights for the genetic algorithm.

An initial population of 50 weights vectors was chosen randomly, with each weight being between –1.0 and +1.0. Montana and Davis tried a number of different genetic operators in various experiments. The mutation and crossover operators they used for their comparison of the genetic algorithm with backpropagation are illustrated in Figures 4.5 and 4.6. The mutation operator selects n non-input units, and for each incoming link to those units, adds a random value between −1.0 and +1.0 to the weight on the link. The crossover operator takes two parent weight vectors, and for each non-input unit in the offspring vector, selects one of the parents at random and copies the weights on the incoming links from that parent to the offspring. Notice that only one offspring is created.

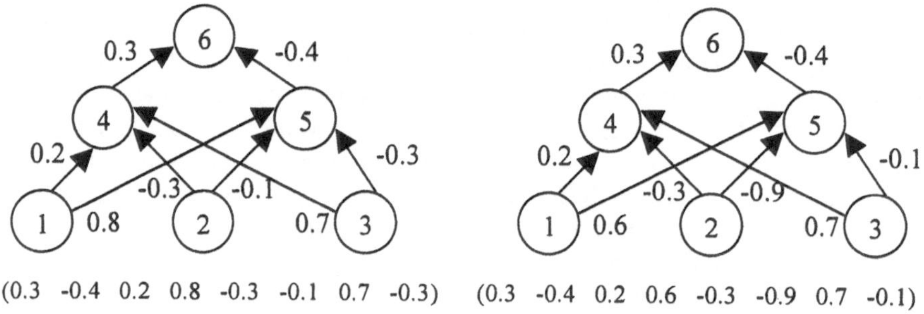

Figure 4.5 Illustration of the mutation method. Here the weights on incoming links to unit 5 are mutated.

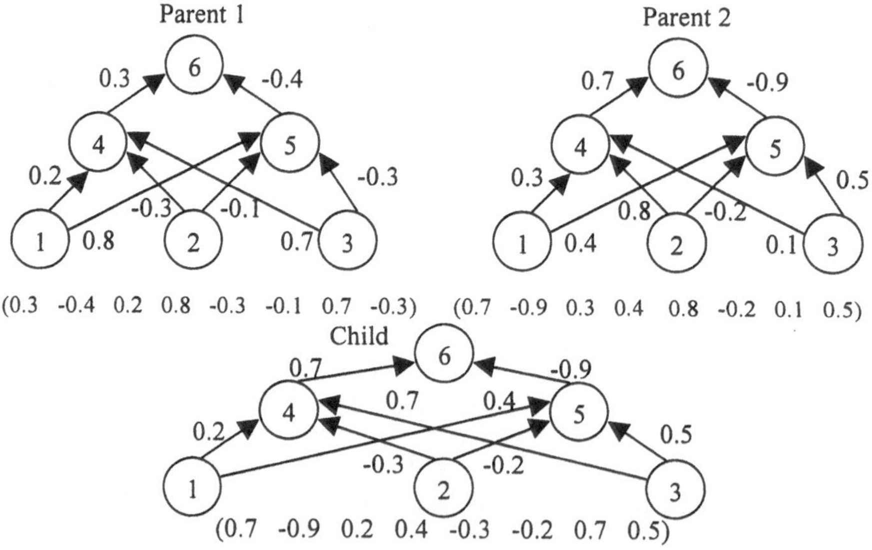

Figure 4.6 Illustration of the crossover method. In the child network shown here, the incoming links to unit 4 come from parent 1 and the incoming links 5 and 6 come from parent 2.

The performance of a genetic algorithm using these operators was compared with the performance of a backpropagation algorithm. The genetic algorithm had a population of 50 weight vectors, and a rank selection method was used. The genetic algorithm was allowed to run for 200 generations. The backpropagation algorithm was allowed to run for 5000 iterations, where one iteration is a complete epoch (a complete pass through the training data). Montana and Davis found that the genetic algorithm significantly outperforms backpropagation on this task, obtaining better weight vectors more quickly.

This experiment shows that in some situations the genetic algorithm is a better training method for neural networks than simple backpropagation. This does not mean that the genetic algorithm will outperform backpropagation in all cases. It is also possible that enhancements of backpropagation might help it overcome some of the problems that prevented it from performing as well as the genetic algorithm in this experiment.

4.3.1.2 Evolving Network Architectures

Neural network researchers know all too well that the particular architecture chosen can determine the success or failure of the application, so they would like very much to be able to automatically optimize the procedure of designing an architecture for a particular application. Many believe that genetic algorithms are well suited for this task (Mitchell, 1996). There have been several efforts along these lines, most of which fall into one of two categories: direct encoding and grammatical encoding. Under direct encoding a network architecture is directly encoded into a genetic algorithm chromosome. Under grammatical encoding, the genetic algorithm does not evolve network architectures; rather, it evolves grammars that can be used to develop network architectures.

Direct Encoding.
The method of direct encoding is illustrated in work done by Geoffrey Miller, Peter Todd, and Shailesh Hedge (1989), who restricted their initial project to feedforward networks with a fixed number of units for which the genetic algorithm was used to evolve the connection topology. As is shown in Figure 4.7, the connection topology was represented by a NxN matrix (5x5 in Figure 4.7) in which each entry encodes the type of connection from the "from unit" to the "to unit". The entries in the connectivity matrix were either "0" (meaning no connection) or "L" (meaning a "learnable" connection). Figure 4.7 also shows how the connectivity matrix was transformed into a chromosome for the genetic algorithm ("0" corresponds to 0 and "L" to 1) and how the bit string was decoded into a network. Connections that were specified to be learnable were initialized with small random weights.

From unit		1	2	3	4	5
To unit	1	0	0	0	0	0
	2	0	0	0	0	0
	3	L	L	0	0	0
	4	L	L	0	0	0
	5	0	0	L	L	0

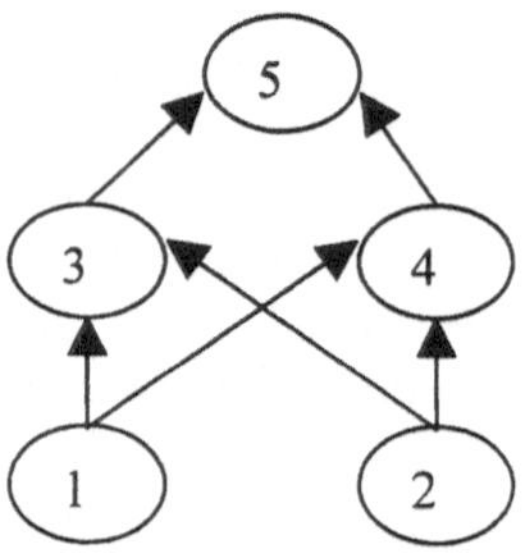

Chromosome: 0 0 0 0 0 0 0 0 0 0 1 1 0 0 0 1 1 0 0 0 0 0 1 1 0

Figure 4.7 Illustration of Miller, Todd, and Hedge's representation scheme.

Miller, Todd, and Hedge used a simple fitness-proportionate selection method and mutation (bits in the string were flipped with some low probability). Their crossover operator randomly chose a row index and swapped the corresponding rows between the two parents to create two offspring. The intuition behind that operator was similar to that behind Montana and Davis's crossover operator-each row represented all the incoming connections to a single unit, and this set was thought to be a functional building block of the network. The fitness of a chromosome was calculated in the same way as in Montana and Davis's project: for a given problem, the network was trained on a training set for a certain number of epochs, using backpropagation to modify the weights. The fitness of the chromosome was the sum of the squares of the errors on the training set at the last epoch. Again, low error translated to high fitness. Miller, Todd, and Hedge tried their genetic algorithm on several problems with very good results. The problems were relatively easy for multilayer neural networks to learn to solve under backpropagation. The networks had different number of units for different tasks; the goal was to see if the genetic algorithm could discover a good connection topology for each task. For each run the population size was 50, the crossover rate was 0.6, and the mutation rate was 0.005. In all cases, the genetic algorithm was easily able to find networks that readily learned to map inputs to outputs over the training set with little error. However, the tasks were too easy to be a rigorous test of this method-it remains to be seen if this method can scale up to more complex tasks that require much larger networks with many more interconnections.

Grammatical Encoding.
The method of grammatical encoding can be illustrated by the work of Hiroaki Kitano (1990), who points out that direct encoding approaches become increasingly difficult to use as the size of the desired network increases. As the

network's size grows, the size of the required chromosome increases quickly, which leads to problems both in performance and in efficiency. In addition, since direct encoding methods explicitly represent each connection in the network, repeated or nested structures cannot be represented efficiently, even though these are common for some problems.

The solution pursued by Kitano and others is to encode networks as grammars; the genetic algorithm evolves the grammars, but the fitness is tested only after a "development" step in which a network develops from the grammar. A grammar is a set of rules that can be applied to produce a set of structures (e.g., sentences in a natural language, programs in a computer language, neural network architectures).

Kitano applied this general idea to the development of neural networks using a type of grammar called a "graph-generation grammar", a simple example of which is given in Figure 4.8a. Here the right-hand side of each rule is a 2x2 matrix rather than a one-dimensional string. Each lower-case letter from a through p represents one of the 16 possible 2x2 arrays of ones and zeros. There is only one structure that can be formed from this grammar: the 8x8 matrix shown in Figure 4.8b. This matrix can be interpreted as a connection matrix for a neural network: a 1 in row i and column i means that unit i is present in the network and a 1 in row i and column , $i \neq j$, means that there is connection from unit i to unit j. The result is the network shown in Figure 4.8c which, with appropriate weights, computes the Boolean function XOR.

Kitano's goal was to have a genetic algorithm evolve such grammars. Figure 4.9 illustrates a chromosome encoding the grammar given in Figure 4.8a. The chromosome is divided up into separate rules, each of which consists of five elements. The first element is the left-hand side of the rule; the second through fifth elements are the four symbols in the matrix on the right-hand side of the rule. The possible values for each element are the symbols A-Z and a-p. The first element of the chromosome is fixed to be the start symbol, S; at least one rule taking S into a 2x2 matrix is necessary to get started in building a network from a grammar

The fitness of a grammar was calculated by constructing a network from the grammar, using backpropagation with a set of training inputs to train the resulting network to perform a simple task, and then, after training, measuring the sum of the squares of the errors made by the network on either the training set or a separate test set. The genetic algorithm used fitness-proportionate selection, multi-point crossover, and mutation. A mutation consisted of replacing one symbol in the chromosome with a randomly chosen symbol from the A-Z and a-p alphabets. Kitano used what he called "adaptive mutation": the probability of mutation of an offspring depended on the Hamming distance (number of mismatches) between the two parents. High distance resulted in low mutation, and vice versa. In this way, the genetic algorithm tended to respond to loss of diversity in the population by selectively raising the mutation rate.

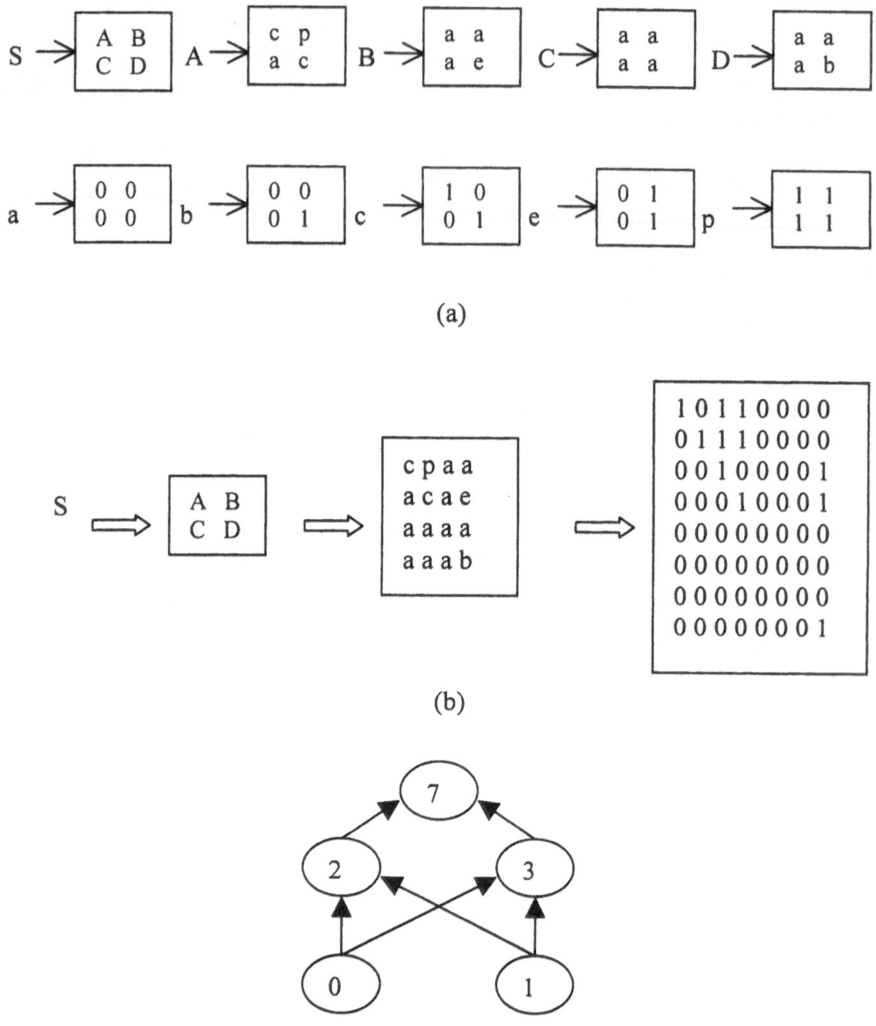

Figure 4.8 Illustration of Kitano's graph generation grammar for the XOR problem. (a) Grammatical rules. (b) A connection matrix is produced from the grammar. (c) The resulting network.

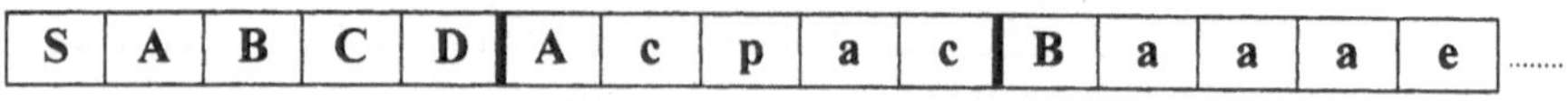

Figure 4.9 Illustration of a chromosome encoding a grammar.

Kitano (1990) performed a series of experiments on evolving networks for simple "encoder/decoder" problems to compare the grammatical and direct encoding approaches. He found that, on these relatively simple problems, the performance of a genetic algorithm using the grammatical encoding method consistently surpassed that of a genetic algorithm using the direct encoding method, both in the correctness of the resulting neural networks and in the speed with which they were found by the genetic algorithm. In the grammatical encoding runs, the genetic algorithm found networks with lower error rate, and found the best networks more quickly, than in direct encoding runs. Kitano also discovered that the performance of the genetic algorithm scaled much better with network size when grammatical encoding was used.

What accounts for the grammatical encoding method's apparent superiority? Kitano argues that the grammatical encoding method can easily create "regular", repeated patterns of connectivity, and that this is a result of the repeated patterns that naturally come from repeatedly applying grammatical rules. We would expect grammatical encoding approaches, to perform well on problems requiring this kind of regularity. Grammatical encoding also has the advantage of requiring shorter chromosomes, since the genetic algorithm works on the instructions for building the network (the grammar) rather that on the network structure itself. For complex networks, the later could be huge and intractable for any search algorithm.

Evolving a Learning Rule.

David Chalmers (1990) took the idea of applying genetic algorithms to neural networks in a different direction: he used genetic algorithms to evolve a good learning rule for neural networks. Chalmers limited his initial study to fully connected feedforward networks with input and output layers only, no hidden layers. In general a learning rule is used during the training procedure for modifying network weights in response to the network's performance on the training data. At each training cycle, one training pair is given to the network, which then produces an output. At this point the learning rule is invoked to modify weights. A learning rule for a single layer, fully connected feedforward network might use the following local information for a given training cycle to modify the weight on the link from input unit i to output unit j:

a_i: the activation of input i

o_j: the activation of output unit j

t_j: the training signal on output unit j

w_{ij}: the current weight on the link from i to j

The change to make in the weight w_{ij} is a function of these values:

$\Delta w_{ij} = f(a_i, o_j, t_j, w_{ij})$.

The chromosomes in the genetic algorithm population encoded such functions.

Chalmers made the assumption that the learning rule should be a linear function of these variables and all their pairwise products. That is, the general form of the learning rule was

$\Delta w_{ij} = k_0(k_1w_{ij}+k_2a_i+k_3o_j+k_4t_j+k_5w_{ij}a_i+k_6w_{ij}o_j+k_7w_{ij}t_j+k_8a_io_j+k_9a_it_j+k_{10}o_jt_j)$.
The k_m ($1<m<10$) are constant coefficients, and k0 is a scale parameter that affects how much the weights can change on any cycle. Chalmer's assumption about the form of the learning rule came in part from the fact that a known good learning rule for such networks is the "delta rule". One goal of Chalmer's work was to see if the genetic algorithm could evolve a rule that performs as well as the delta rule.

The task of the genetic algorithm was to evolve values for the km's. The chromosome encoding for the set of km's is illustrated in Figure 4.10. The scale parameter k_0 is encoded as five bits, with the zeroth bit encoding the sign (1 encoding + and 0 encoding -) and the first through fourth bits encoding an integer n: $k_0 = 0$ if $n = 0$; otherwise $|k_0| = 2^{n-9}$. Thus k_0 can take on the values 0,+-1/256,+-1/128,...,+-32,+-64. The other coefficients km are encoded by three bits each, with the zeroth bit encoding the sign and the first and second bits encoding an integer n. For i=1,...,10, km=0 if n=0; otherwise $|k_m| = 2^{n-1}$.

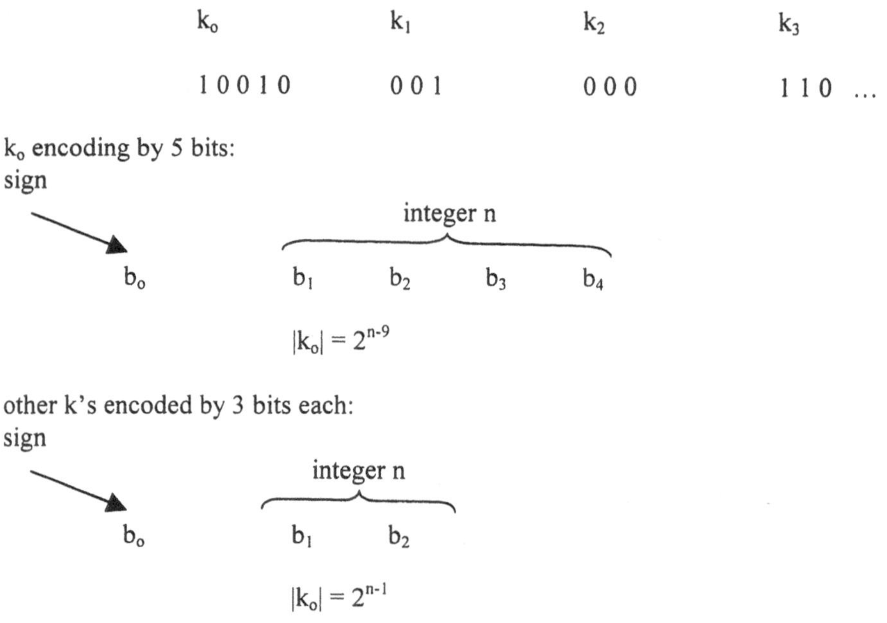

Figure 4.10 Illustration of the method for encoding the km's.

The fitness of each chromosome (learning rule) was determined as follows. A subset of 20 mappings was selected from the full set of 30 linear separable mappings. For each mapping, 12 training examples were selected. For each of these mappings, a network was created with the appropriate number of input units for the given mapping. The network's weights were initialized randomly. The network was run on the training set for some number of epochs (typically 10), using the learning rule specified by the chromosome. The performance of the learning rule on a given mapping was a function of the network's error on the training set, with low error meaning high performance. The overall fitness of the learning rule was a function of the average error of 20 networks over the chosen subset of 20 mappings. This fitness was then transformed to be a percentage, where a high percentage meant high fitness.

Using this fitness measure, the genetic algorithm was run on a population of 40 learning rules, with two-point crossover and standard mutation. The crossover rate was 0.8 and the mutation rate was 0.01. Typically, over 1000 generations, the fitness of the best learning rules in the population rose from between 40% and 60% in the initial generation to between 80% and 98%. The fitness of the delta rule is around 98%, and on one out of a total of ten run the genetic algorithm discovered this rule. On three of the ten runs, the genetic algorithm discovered slight variations of this rule with lower fitness.

These results show that, given a somewhat constrained representation, the genetic algorithm was able to evolve a successful learning rule for simple single layer networks. The extent to which this method can find learning rules for more complex networks remains an open question, but these results are a first step in that direction. Chalmers suggested that it is unlikely that evolutionary methods will discover learning methods that are more powerful than backpropagation, but he speculated that genetic algorithms might be a powerful method for discovering learning rules for unsupervised learning paradigms or for new classes of network architectures.

4.3.2 Evolving Fuzzy Systems

Ever since the very first introduction of the fundamental concept of fuzzy logic by Zadeh in 1973, its use in engineering disciplines has been widely studied. Its main attraction undoubtedly lies in the unique characteristics that fuzzy logic systems possess. They are capable of handling complex, non-linear dynamic systems using simple solutions. Very often, fuzzy systems provide a better performance than conventional non-fuzzy approaches with less development cost.

However, to obtain an optimal set of fuzzy membership functions and rules is not an easy task. It requires time, experience, and skills of the operator for the tedious fuzzy tuning exercise. In principle, there is no general rule or method for the fuzzy logic set-up. Recently, many researchers have considered a number of intelligent techniques for the task of tuning the fuzzy set.

Here, another innovative scheme is described (Man, Tang & Kwong, 1999). This approach has the ability to reach an optimal set of membership functions and rules without a known overall fuzzy set topology. The conceptual idea of this approach is to have an automatic and intelligent scheme to tune the membership functions and rules, in which the conventional closed loop fuzzy control strategy remains unchanged, as indicated in Figure 4.11.

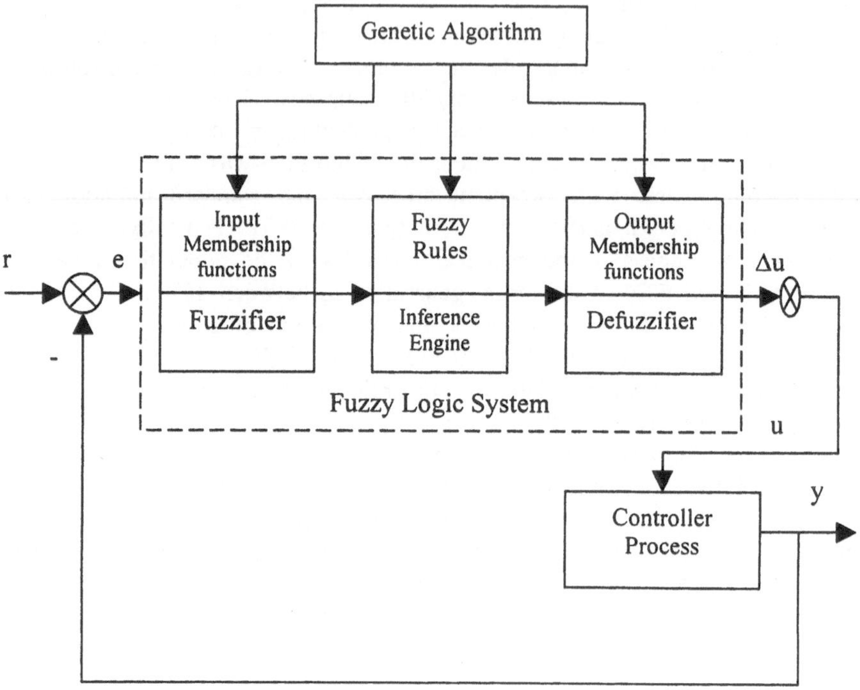

Figure 4.11 Genetic algorithm for a fuzzy control system.

In this case, the chromosome of a particular is shown in Figure 4.12. The chromosome consist of two types of genes, the control genes and parameter genes. The control genes, in the form of bits, determine the membership function activation, whereas the parameter genes are in the form of real numbers to represent the membership functions.

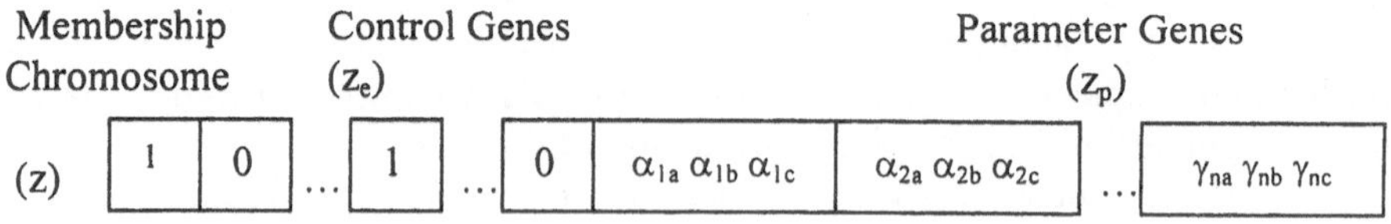

Figure 4.12 Chromosome structure for the fuzzy system.

To obtain a complete design for the fuzzy control system, an appropriate set of fuzzy rules is required to ensure system performance. At this point it should be stressed that the introduction of the control genes is done to govern the number of fuzzy subsets in the system.

Once the formulation of the chromosome has been set for the fuzzy membership functions and rules, the genetic operation cycle can be performed. This cycle of operation for the fuzzy control system optimization using a genetic algorithm is illustrated in Figure 4.13.

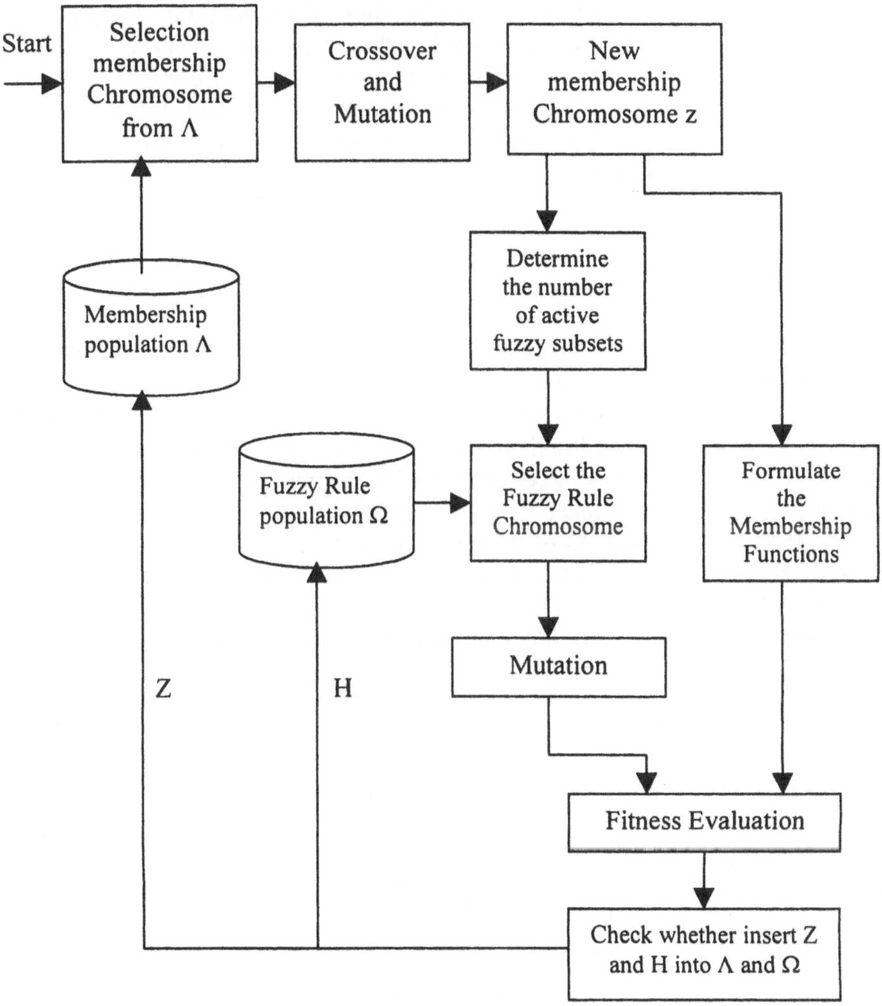

Figure 4.13 Genetic cycle for fuzzy system optimization.

There are two population pools, one for storing the membership chromosomes and the other for storing the fuzzy rule chromosomes. We can see this in Figure 4.13 as the membership population and fuzzy rule population, respectively. Considering that there are various types of gene structure, a number of different genetic operations can be used. For the crossover operation, a one point crossover is applied separately for both the control and parameter genes of the membership chromosomes within certain operation rates. There is no crossover operation for fuzzy rule chromosomes since only one suitable rule set can be assisted.

Bit mutation is applied for the control genes of the membership chromosome. Each bit of the control gene is flipped if a probability test is satisfied (a randomly generated number is smaller than a predefined rate). As for the parameter genes, which are real number represented, random mutation is applied.

The complete genetic cycle continues until some termination criteria, for example, meeting the design specification or number of generation reaching a predefined value, are fulfilled.

4.4 Summary

We have presented in this chapter the basic concepts of genetic algorithms and simulated annealing. These optimization methodologies are motivated by nature's wisdom. Genetic algorithms emulate the process of evolution in nature, while simulated annealing is based in thermodynamic processes. We have also presented the application of genetic algorithms to the problems of optimizing neural networks and fuzzy systems. Genetic algorithms can then be viewed as a technique for efficient design of intelligent systems, because they can be used to optimize the weights or architecture of the neural network, or the number of rules in a fuzzy system. In later chapter we will make use of this fact to design intelligent systems for controlling real world dynamical systems.

Chapter 5

Dynamical Systems Theory

This chapter introduces the basic concepts of dynamical systems theory, and several basic mathematical methods for controlling chaos. The main goal of this chapter is to provide an introduction to and a summary to the theory of dynamical systems with particular emphasis on fractal theory, chaos theory, and chaos control. We first define what is meant by a dynamical system, then we define an attractor, and the concept of the fractal dimension of a geometrical object. Also, we define the Lyapunov exponents as a measure of the chaotic behavior of a dynamical system. On the other hand, the fractal dimension can be used to classify geometrical objects because it measures the complexity of an object. We finish the chapter by reviewing mathematical methods for controlling chaos in dynamic systems. These methods can be used to control a real dynamic system, however, due to efficiency and accuracy requirement we will be forced to consider more advanced methods in the following chapters.

5.1 Basic Concepts of Dynamical Systems

In this section we present a brief overview of the field of Non-Linear Dynamical Systems and Fractal Theory. Recently research has shown that many simple non-linear deterministic systems can behave in an apparently unpredictable and "chaotic" manner (Grebogi, Ott, & Yorke, 1987). The existence of complicated dynamics has been discussed in the mathematical literature for many decades with important contributions by Poincaré, Birkhoft, Smale and Kolmogorov and his students, among others. Nevertheless, it is only recently that the wide-ranging impact of "chaos" has been recognized. Consequently, the field is now undergoing explosive growth, and many applications have been made across a broad spectrum of scientific disciplines-robotics, engineering, physics, chemistry, fluid mechanics and economics, to name several. We start with some basic definitions of concepts used in this book.

Dynamic System: This is a set of mathematical equations that allows one, in principle, to predict the future behavior of the system given the past. One example is a system of first-order ordinary differential equations in time:

$$\frac{dx}{dt} = G(x,t) \tag{5.1}$$

where x(t) is a D-dimensional vector and G is a D-dimensional vector function of x and t. Another example is a map.

Map: A map is an equation of the following form:

$$x_{t+1} = F(x_t) \tag{5.2}$$

where the "time" t is discrete and integer valued. Thus, given x_o, the maps gives x_1. Given x_1, the map gives x_2, and so on.

Dissipative system: In Hamiltonian (conservative) systems such as the ones arising in Newtonian mechanics of particles (without friction), phase space volumes are preserved by time evolution (the phase space is the space of variables that specify the state of the system). Consider, for example, a two-dimensional phase space (q, p), where q denotes a position variable and p a momentum variable. Hamilton's equations of motion take the set of initial conditions at time t $=t_o$ and evolve them in time to the set at time $t = t_1$. Although the shapes of the sets are different, their areas are the same. By a dissipative system we mean one that does not have this property. Areas should typically decrease (dissipate) in time so that the area of the final set would be less than the area of the initial set. As a consequence of this, dissipative systems typically are characterized by the presence of attractors.

Attractor: If one considers a system and its phase space, then the initial conditions may be attracted to some subset of the phase space (the attractor) as time $t \rightarrow \infty$. For example, for a damped harmonic oscillator the attractor is the point at rest. For a periodically driven oscillator in its limit cycle the limit set is a closed curve in the phase space.

Strange attractor: In the above two examples, the attractors were a point, which is a set of dimension zero, and closed curve, which is a set of dimension one. For many other attractors the attracting set can be much more irregular (some would say pathological) and, in fact, can have a dimension that is not an integer. Such sets have been called "fractal" and, when they are attractors, they are called strange attractors. The existence of a strange attractor in a physically interesting model was first demonstrated by Lorenz (see Lorenz, 1963).

Chaotic attractor: By this term we mean that if we take two typical points on the attractor that are separated from each other by a small distance $\Delta(0)$ at $t = 0$, then for increasing t they move apart exponentially fast. That is, in some average sense:

$$\Delta(t) \sim \Delta(0) \exp(\lambda\, t) \tag{5.3}$$

with $\lambda > 0$ (where λ is called the Lyapunov exponent). Thus a small uncertainty in the initial state of the system rapidly leads to inability to forecast its future. It is typically the case that strange attractors are also chaotic.

One of the most prominent, chaotic, continuous-time dynamical systems is the "Lorenz attractor", named after the meteorologist E.N. Lorenz who investigated the three-dimensional, continuous-time system

$$\begin{aligned} x' &= S(-x + y) \\ y' &= rx - y - xz \qquad s, r, b > 0 \\ z' &= -bz + xy \end{aligned} \tag{5.4}$$

emerging in the study of turbulence in fluids. For r above the critical value of r = 28.0, trajectories of Equation (5.4) evolve in a rather unexpected way. Suppose that a trajectory starts at an initial value near the origin. For some time the trajectory regularly spirals outward from one fixed point, then the trajectory jumps to a region near another fixed point and does the same thing. As trajectories starting at different initial values all converge to and remain in the same region near the two fixed points, the region is considered an "attractor". It is a "strange attractor" because it is neither a point nor a closed curve. In general, this chaotic behavior can only occur for systems of at least three simultaneous non-linear differential equations or for systems of at least a one-dimensional non-linear map (Devaney, 1989).

Fractal geometry is a mathematical tool for dealing with complex systems that have no characteristic length scale. A well known example is the shape of a coastline. When we see two pictures of a coastline on two different scales, we cannot tell which scale belongs to which picture: both look the same. This means that the coastline is scale invariant or, equivalently, has no characteristic length scale. Other examples in nature are rivers, cracks, mountains, and clouds. Scale-invariant systems are usually characterized by non-integer ("fractal") dimensions.

The dimension tells us how some property of an object or space changes as we view it at increased detail. There are several different types of dimension. The fractal dimension d_f describes the space filling properties of an object. Three examples of the fractal dimension are the self-similarity dimension, the capacity dimension, and the Hausdorff-Besicovitch dimension. The topological dimension d_T describes how points within an object are connected together. The embedding dimension d_e describes the space in which the object is contained.

The fractal dimensions d_f are useful and important tools to quantify self-similarity and scaling. Essentially, the dimension tells us how many new pieces are resolved as the resolution is increased. The self-similarity dimension can only be applied to geometrical self-similar objects, where the small pieces are exact copies of the whole object. However the capacity dimension can be used to analyze irregularly shaped objects that are statistically self-similar. On the other hand, the Hausdorff-Besicovitch dimension requires more complex mathematical tools. For this reason, we will limit our discussion here to the capacity dimension.

A ball is the set of points within radius r of a given point. We determine N(r) the minimum number of balls required so that each point in the object is contained within at least one ball of radius r. In order to cover all the points of the object, the balls may need to overlap. The capacity dimension is defined by the following equation:

$$d_c = \lim_{r \to 0} \frac{\log N(r)}{\log(1/r)} \quad . \tag{5.5}$$

The capacity dimension defined as above is a measure of the space filling properties of an object because it gives us an idea of how much work is needed to cover the object with balls of changing size.

A useful method to determine the capacity dimension is to choose balls that are the non-overlapping boxes of a rectangular coordinate grid. N(r) is then the number of boxes with side of length r that contain at least one point of the object. Efficient algorithms have been developed to perform this "box counting" for different values of r, and thus determine the box counting dimension as the best fit of log N(r) versus log(1/r).

The fractal dimension d_f characterizes the space-filling properties of an object. The topological dimension d_T characterizes how the points that make up the object are connected together. It can have only integer values. Consider a line that is so long and wiggly that it touches every point in a plane and thus covers an area. Because it covers a plane, its space-filling fractal dimension $d_f = 2$. However, no matter how wiggly it is, it is still a line and thus has topological dimension $d_T=1$. Thus, the essence of a fractal is that its space-filling properties are larger than one anticipates from its topological dimension. Thus we can now present a formal definition of a fractal (Mandelbrot, 1987), namely, that an object is a fractal if and only if

$$d_f > d_T \ .$$

However, there is no one definition that includes all the objects or processes that have fractal properties.

Despite the identification of fractals in nearly every branch of science, too frequently the recognition of fractal structure is not accompanied with any additional insight as to its cause. Often we do not even have the foggiest idea as to the underlying dynamics leading to the fractal structure. The chaotic dynamics of non-linear systems, on the other hand, is one area where considerable progress has been made in understanding the connection with fractal geometry. Indeed, chaotic dynamics and fractal geometry have such a close relationship that one of the hallmarks of chaotic behavior has been the manifestation of fractal geometry, particularly for strange attractors in dissipative systems (Rasband, 1990). For a practical definition we take a "strange attractor", for a dynamic system, to be an attracting set with fractal dimension. For example, the famous Lorenz strange attractor has a fractal dimension of about 2.06. Also, we think that beyond only this relationship between strange attractors and the fractal dimension of the set, there is a deeper relationship between the underlying dynamics of a system and the fractal nature of its behavior. We will explore this relationship in more detail in the following chapter.

Let us consider as an example, the use of the fractal dimension as a mathematical model of the time series in the following form:

$$d = [\log(N)/\log(1/r)] \tag{5.6}$$

where d is the fractal dimension for an object of N parts, each scaled down by a ratio r. For an estimation of this dimension we can use the following equation:

$$N(r) = \beta[\ 1/r^d\ ] \tag{5.7}$$

where N(r) = number of boxes contained in a geometrical object and r = size of the box. We can obtain the box dimension of a geometrical object (Mandelbrot, 1987) counting the number of boxes for different sizes and performing a logarithmic regression on this data. For our particular case the geometrical object consists of the curve constructed using the set of points from the time series. We show in Figure 5.1 (a) the curve and the boxes used to cover it. In Figure 5.1 (b) the corresponding logarithmic regression is illustrated.

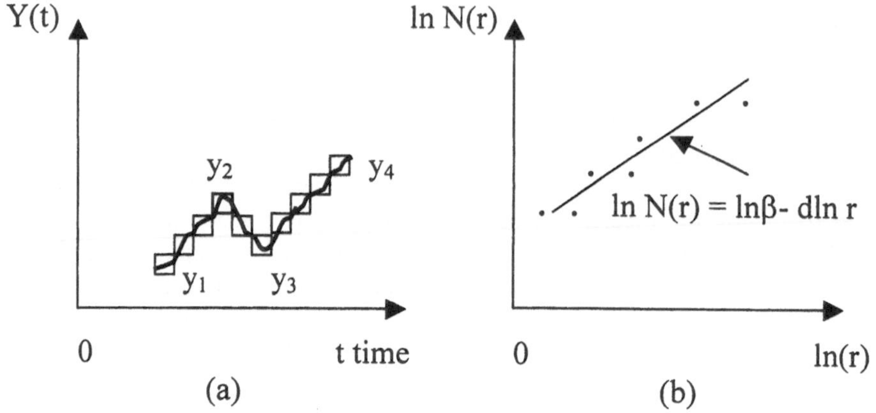

Figure 5.1 Fractal dimension of a time series: (a) curve and the boxes covering it, (b) the logarithmic regression to find d

5.2 Controlling Chaos

More than two decades of intensive studies on non-linear dynamics have posed the question on the practical applications of chaos (Kapitaniak, 1996). One of the possible answers is to control chaotic behavior in such a way as to make it predictable. Indeed, nowadays the idea of controlling chaos is an appealing one.

Chaos occurs widely in engineering and natural systems; historically it has usually been regarded as a nuisance and is designed out if possible. It has been noted only as irregular or unpredictable behavior, often attributed to random external influences. More recently, there have been examples of the potential usefulness of chaotic behavior (Kapitaniak, 1996).

We can divide chaos controlling approaches into two broad categories: firstly those in which the actual trajectory in the phase space of the system is monitored and some "feedback" process is employed to maintain the trajectory in the desired mode, and secondly "non-feedback" methods in which some other property or knowledge of the system is used to modify or exploit chaotic behavior. Feedback methods do not change the controlled systems and stabilize unstable periodic orbits or strange chaotic attractors, while non-feedback methods slightly change the controlled system, mainly by a small permanent shift of control parameter, changing the system behavior from chaotic attractor to periodic orbit which is close to the initial attractor.

We describe in this section several methods by which chaotic behavior in a dynamical system may be modified, displaced in parameter space or removed. The Ott-Grebogi-Yorke (OGY) method (Ott et al.,1990) is extremely general, relying only on the universal property of chaotic attractors, namely that they have embedded within them infinitely many unstable periodic orbits. On the other hand, the method requires following the trajectory and employing a feedback control system, which must be highly flexible and responsive; such a system in some experimental configurations may be large and expensive. It has the additional disadvantage that small amounts of noise may cause occasional large departures from the desired operating trajectory.

The non-feedback approach is inevitable much less flexible, and requires more prior knowledge of equations of motion. On the other hand, to apply such a method, we do not have to follow the trajectory. The control procedures can be applied at any time and we can switch from one periodic orbit to another without returning to the chaotic behavior, although after each switch, transient chaos may be observed. The lifetime of this transient chaos strongly depends on initial conditions. Moreover, in a non-feedback method we do not have to wait until the trajectory is close to an appropriate unstable orbit; in some cases this time can be quit long. The dynamic approach can be very useful in mechanical systems, where feedback controllers are often very large. In contrast, a dynamical absorber having a mass of the order of 1% of that of the control system is able, as we will see later, to convert chaotic behavior to periodic one over a substantial region of parameter space. Indeed, the simplicity by which chaotic behavior may be changed in this way may actually motivate the search for, and exploitation of, chaotic behavior in practical systems.

The essential property of a chaotic trajectory is that it is not asymptotically stable.
Closely correlated initial conditions have trajectories, which quickly become uncorrelated. Despite this obvious disadvantage, it has been established that control leading to the synchronization of two chaotic systems is possible.

The methods described in this section are illustrated by the example of Chua's circuit (Chua, 1993) shown in Figure 5.2. Chua's circuit contains three linear energy storage elements (an inductor and two capacitors), a linear resistor, and a single non-linear resistor NR, namely Chua's diode with a three segment piecewise linear v-i characteristic defined by

$$f(v_{c1})=m_0 v_{c1} + \tfrac{1}{2}(m_1 - m_0)(|v_{c1} + 1|-| v_{c1} - 1|) \qquad (5.8)$$

where the slopes in the inner and outer regions are m0 and m1 respectively (Figure 5.3).

In this case the state equations for the dynamics of Chua's circuit are as follows:

$$C_1\frac{dv_{c1}}{dt} = G(v_{c2} - v_{c1}) - f(v_{c1})$$
$$C_2\frac{dv_{c2}}{dt} = G(v_{c1} - v_{c2}) - i_L \qquad (5.9)$$
$$L\frac{di_L}{dt} = v_{c2}$$

where G=1/R.

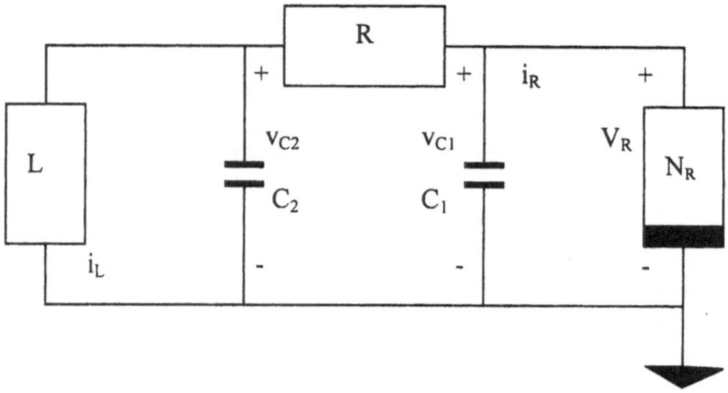

Figure 5.2 Chua's circuit.

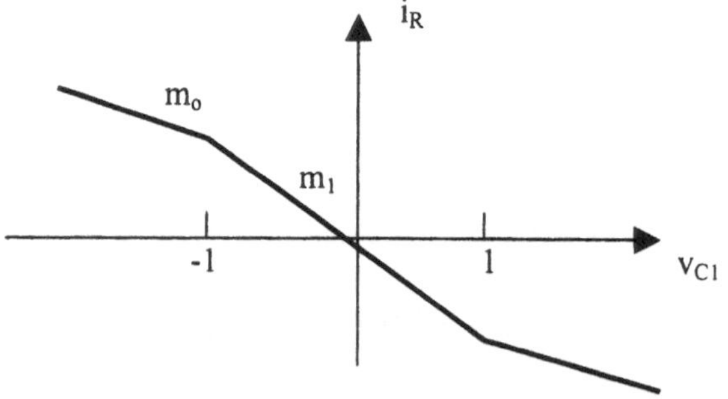

Figure 5.3 i_R-V_{c1} characteristic of the non-linear resistor.

The chaotic dynamics of Chua's circuit have been widely investigated (e.g. Madan, 1993). One of the main advantages of this system is the very good accuracy between numerical simulations of the model and experiments on real electronic devices. Experiments with this circuit are very easy to perform, even for non-specialists.

5.2.1 Controlling Chaos through Feedback

5.2.1.1 Ott-Grebogi-Yorke Method

Ott, Grebogi and Yorke (Ott et al., 1990) have proposed an developed a method by which chaos can always be suppressed by shadowing one of the infinitely many unstable periodic orbits embedded in the chaotic attractor.

The basic assumptions of this method are as follows.

(a) The dynamics of the system can be described by an n-dimensional map of the form:

$$X_{n+1} = f(X_n, p) \tag{5.10}$$

(b) p is some accessible system parameter which can be changed in some small neighborhood of its nominal value p*.
(c) For this value p* there is a periodic orbit within the attractor around which we would like to stabilize the system.
(d) The position of this orbit changes smoothly with changes in p, and there are small changes in the local system behavior for small variations of p.

Let X_F be a chosen fixed point of the map f of the system existing for the parameter value p*. In the close vicinity of this fixed point with good accuracy we can assume that the dynamics are linear and can be expressed approximately by

$$X_{n+1} - X_F = M(X_n - X_F) \tag{5.11}$$

The elements of the matrix M can be calculated using the measured chaotic time series and analyzing its behavior in the neighborhood of the fixed point. The OGY algorithm is schematically explained in Figure 5.4 and its main properties are as follows.

(a) No model of dynamics is required. One can use either full information from the process or a delay coordinate embedding technique using single variable experimental time series.

(b) Any accessible variable (controllable) system parameter can be used as the control parameter.
(c) In the absence of noise and error, the amplitude of applied control signal must be large enough (exceed a threshold) to achieve control.
(d) Inevitable noise can destabilize the controlled orbit, resulting in occasional chaotic bursts.
(e) Before settling into the desired periodic mode, the trajectory exhibits chaotic transients, the length of which depends on the actual starting point.

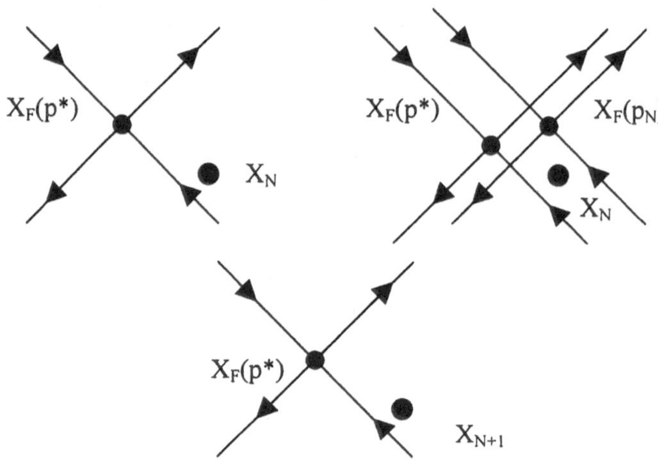

Figure 5.4 General idea of the Ott-Grebogi-Yorke method.

In Ogorzalek (1993) the OGY method was applied to control chaos in Chua's circuit. Using a specific software package, unstable periodic orbits embedded in the attractor which could serve as goals of control were found. The controlling method was implemented in the way shown in Figure 5.5.

The computer was used for data acquisition, identification of the chaotic system in terms of unstable periodic orbits and calculation of the control signal. When applying the OGY method to control chaos in a real electronic circuit the main problem encountered was the noise introduced due to inevitable noise of the circuit elements. The method was found to be very sensitive to the noise level – very small signals sometimes are hidden within the noise, and control is impossible.

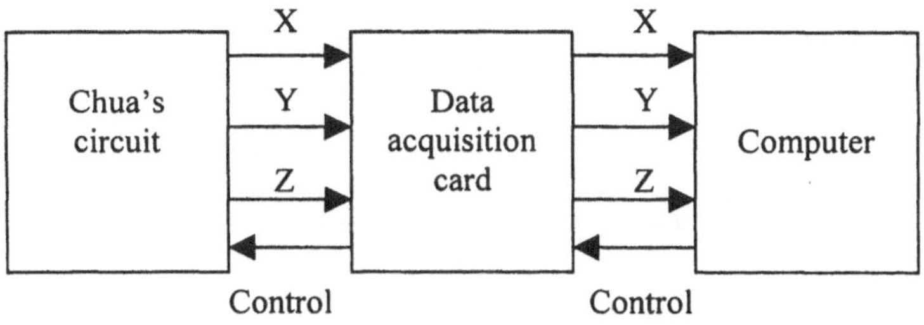

Figure 5.5 Practical implementation of OGY method.

Generally, the experimental application of the OGY method requires a permanent computer analysis of the state of the system. The changes of the parameters, however, are discrete in time and this leads to some serious limitations. The method can stabilize only those periodic orbits which maximal Lyapunov exponent is small compared to the reciprocal of the time interval between parameter changes. Since the corrections of the parameter are rare and small, the fluctuation noise leads to occasional bursts of the system into regions far from the desired periodic orbit, especially in the presence of noise.

5.2.1.2 Pyragas's Control Methods

A different approach to feedback control which allows the above mentioned problems to be avoided was proposed by Pyragas (1992). This method is based on the construction of a special form of a time continuous perturbation, which does not change the form of the desired unstable periodic orbit, but under certain constraints can stabilize it. Two feedback controlling loops, shown in Figure 5.6, have been proposed.

A combination of feedback and periodic external force is used in the first method (Figure 5.6(a)). The second method (Figure 5.6(b)) does not require any external source of energy and it is based on self-controlling delayed feedback. If the period of external force or a time delay is equal to the period of one of the unstable periodic orbits embedded in the chaotic attractor it is possible to find a constant K which allows stabilization of the unstable periodic orbit. This approach, being noise resistant, can easily be used in experimental systems.

The first of Pyragas's methods can be considered as the special case of the direct application of classical controlling methods to the problem of controlling chaos.

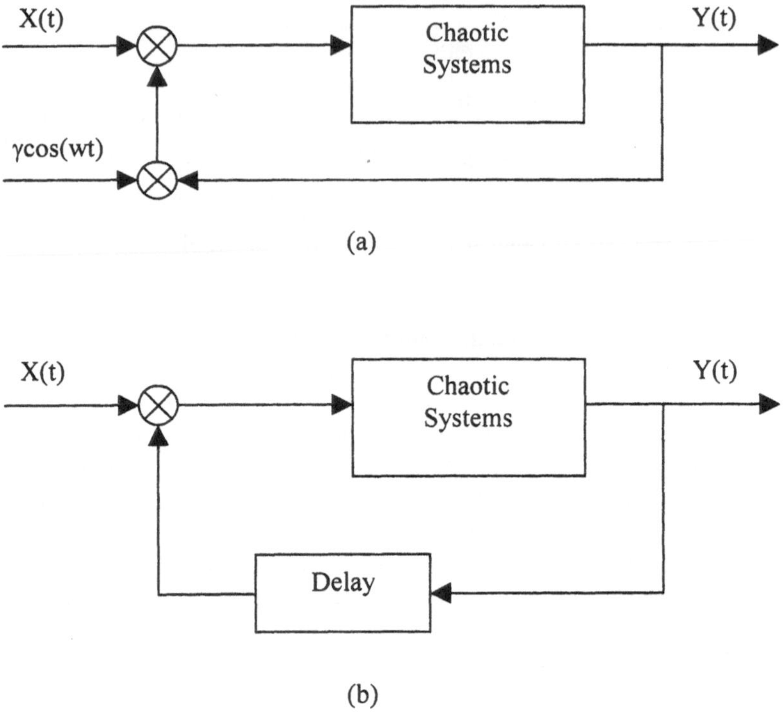

Figure 5.6 Feedback controlling loops; (a) control by periodic external perturbation, (b) control by time delay.

The dynamical system

$$X' = f(X) \tag{5.12}$$

Where $x \in R^n$, is controllable if there exists a control function u(t), such that

$$X' = f(X) + u(t) \tag{5.13}$$

allows to move trajectory X(t) from point X_0 at time t_0 to the desired point X in finite time T.

The controllability concept can be applied to the chaos controlling problems. For example, for Chua's circuit the equations for the controlled circuit are

$$X' = a(Y-X-f(X))$$
$$Y' = X-Y+Z-K(Y-Y^*) \qquad (5.14)$$
$$Z' = -bY$$

This approach is illustrated in Figure 5.7. The main advantages of this method are as follows.

(a) Any solution of the original system can be a goal of the control (fixed point, unstable periodic orbit, etc.)
(b) The controller has a very simple structure.
(c) Access to system parameters is not required.
(d) It is not affected by small parameter variations.

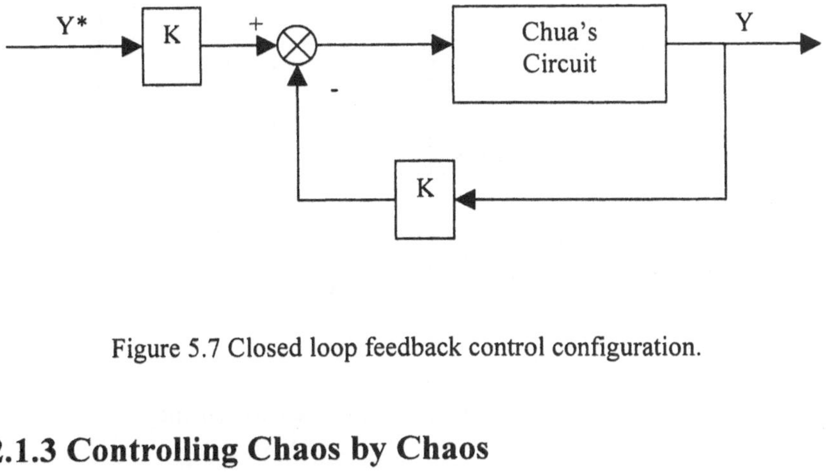

Figure 5.7 Closed loop feedback control configuration.

5.2.1.3 Controlling Chaos by Chaos

In this section, we show that the chaotic behavior of one system can be controlled by coupling it with another one which can also be chaotic (Kapitaniak, 1996). Thus we consider two chaotic systems, which we call A and B respectively,

$$X' = f(X)$$
$$Y' = g(X) \qquad (5.15)$$

where $x,y \in R^n$, and we use the controlling strategy which is schematically illustrated in Figure 5.8; the two systems are coupled through the operators λ,μ, which have a very simple linear form. We assume that some or all state variables of both systems A and B can be measured, so that we can measure signal X(t) from system A and signal Y(t) from B, and that the systems are coupled in such a way that the differences D_1 and D_2 between the signals X(t) and Y(t) are

$$F_1(t) = \lambda[X(t)-Y(t)] = \lambda D_1(t)$$
$$F_2(t) = \mu[Y(t)-X(t)] = \mu D_2(t) \tag{5.16}$$

used as control signals introduced respectively into each of the chaotic systems A and B as negative feedback. We take λ, $\mu > 0$ to be experimentally adjustable weights of the perturbation.

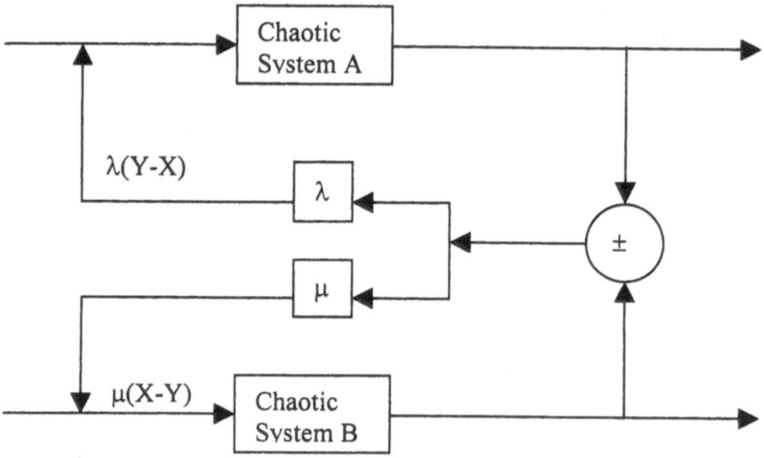

Figure 5.8 Controlling chaos by chaos scheme.

Using the coupling schematically shown in Figure 5.8, it has been shown that one chaotic system coupled with the other one can significantly change the behavior of one of them (unidirectional coupling, i.e. λ or μ =0) or of both systems (mutual coupling, i.e. λ, $\mu \neq 0$). This property allows us to describe the above procedure as the "controlling chaos by chaos" method. In Kocarev and Kapitaniak (1995), rigorous conditions are given, under which chaotic attractors of systems A and B are equivalent, or the evolution of one of them is forced to take place on the attractor of the other one. Kapitaniak (1996) shows an example of coupling two Lorenz chaotic attractors, which results in chaos control and increase of the predictability.

5.2.2 Controlling Chaos without Feedback

5.2.2.1 Control through Operating Conditions

Virtually all engineering and most natural systems are subjected during operation to external forcing. This forcing will contain (and hopefully be dominated by) planned and intentional components; it will also almost invariably contain unintentional "noise". Smart design and control of this forcing is often able to annihilate, or shift to a harmless region of parameter space, an unwanted chaotic behavior.

In this case, the method consists in finding the chaotic region in parameter space by analytical and numerical methods (Kapitaniak, 1996). Then based on this region change the parameters to control the dynamical system. The procedure described in this section is based on the direct change of one of the system parameters to shift system behavior from chaotic to periodic, close to the chaotic attractor. It cannot be called a control method in the sense of the methods described before, but it illustrates that having a system designed as chaotic, we obtain easy access to different types of periodic behavior.

5.2.2.2 Control by System Design

In this section, we explore the idea of modifying or removing chaotic behavior by appropriate system design. It is clear that, to a certain extent, chaos may be "designed out" of a system by appropriate modification of parameters, perhaps corresponding to modification of mass or inertia of moving parts. Equally clearly, there exist strict limits beyond which such modifications cannot go without seriously affecting the efficiency of the system itself.

In this section, we describe a method for controlling chaos in which the chaos effect is achieved by coupling the chaotic main system to a simpler autonomous system (controller), usually linear, as shown in Figure 5.9. This method (Kapitaniak, 1996) is developed for chaotic systems in which for some reason it is difficult, if not impossible, to change any parameter of the main system. In particular consider the coupling of the chaotic system

$$X' = f(X,\mu) \tag{5.17}$$

where $x \in R^n$, $n \geq 3$ and $\mu \in R$ is a system parameter, to another (simpler) asymptotically stable system (controller) described by

$$Y' = g(Y,e) \tag{5.18}$$

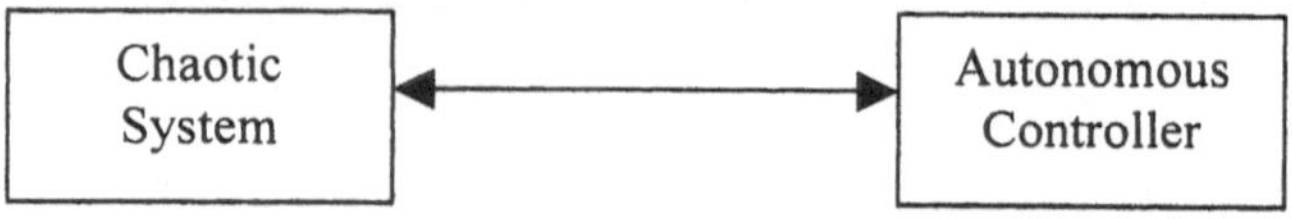

Figure 5.9 Coupling scheme.

where $Y \in R^n$, e is a vector denoting the controller's parameters, where at least one of the parameters e_i can be easily changed. For practical reasons, the dimension m of the controller system (5.18) should be chosen as low as possible. Since the method was mainly designed for controlling chaos in mechanical systems, we choose m = 2, i.e., a one degree of freedom controller (the simplest mechanical system). The equations for the extended system are

$$
\begin{aligned}
X' &= f(X,\mu) + AY \\
Y' &= g(Y,e) + BX
\end{aligned}
\tag{5.19}
$$

where A and B are the coupling matrices. Since the Y subsystem is asymptotically stable, the role of the controller is to change the behavior of the system from chaotic to some desired periodic, possibly constant, operating regime.

The idea of this method is similar to that of the so-called dynamical vibration absorber. A dynamical vibration absorber is a one degree of freedom system, usually mass on a spring, which is connected to the main system as shown in Figure 5.10.

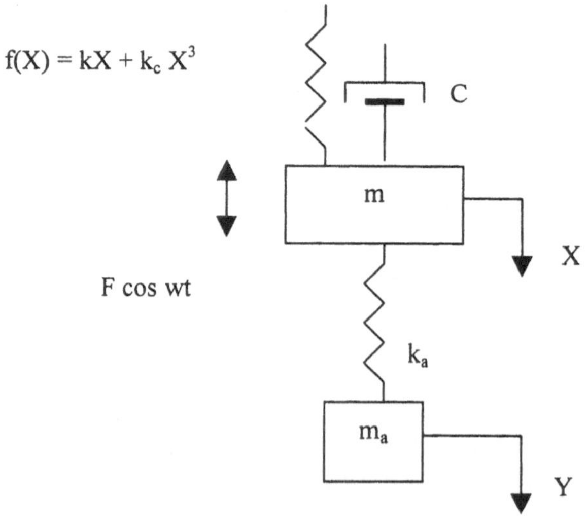

Figure 5.10 Dynamical damper as chaos controller.

Although such a dynamical absorber can change the overall dynamics substantially, it need usually only be physically small in comparison with the main system, and does not require an increase of excitation force. It can be easily added to the existing system without major changes of design or construction. This contrasts with devices based on feedback control, which can be large and costly.

To explain the role of dynamical absorbers in controlling chaotic behavior let us consider the Duffing oscillator, coupled with an additional linear system:

$$X'' + aX' + bX + cX^3 + d(X-Y) = B_0 + B_1\cos wt \qquad (20a)$$

$$Y'' + e(Y-X) = 0 \qquad (20b)$$

where a, b, c, d, e, B_0, B_1, and w are constants. Here d and e are the characteristic parameters for the absorber, and we take e as the control parameter.

It is well known that the Duffing oscillator shows chaotic behavior for certain parameter regions. As has been mentioned in the previous section, in many cases the route to chaos proceeds via s sequence of period doubling bifurcations, and in such cases this method provides an easy way of switching between chaotic and periodic behavior.

Let us consider the parameters of Equation (20) to be fixed at the values a=0.077, b=0, c=1.0, B_0=0.045, B_1=0.16, w=1.0, then we can find (Kapitaniak, 1996) that we have chaos for $e \in [0, 0.10]$, and we can control this chaos by increasing e above 0.10. As this method is designed mainly for experimental applications, we shall now briefly suggest some guidelines for applying it.

(1) The coupled system has to be as simple as possible.
(2) The coupling e should be chosen as small as possible.
(3) If it is possible one should couple the controller in such a way that the locations of the fixed points of the original system are not changed.

5.2.2.3 Taming Chaos

In paper by Steeb et al. (1986) it was first demonstrated that chaos in a dynamical system can be reduced (the largest Lyapunov exponent is decreased) or replaced by regular behavior by applying a weak external periodic signal. Periodic perturbation can be introduced to the system as external force or as a perturbation of one of the internal system parameters. Given an external perturbation, it is possible to show that a chaotic system is capable of finding an appropriate orbit.

In Kapitaniak (1988) it was shown that the chaotic system can be set into regular motion by addition of suitable random noise. Other authors have described similar approaches. Later these approaches have been called "taming chaos". Possible outputs of the taming chaos procedure are shown in Figure 5.11.

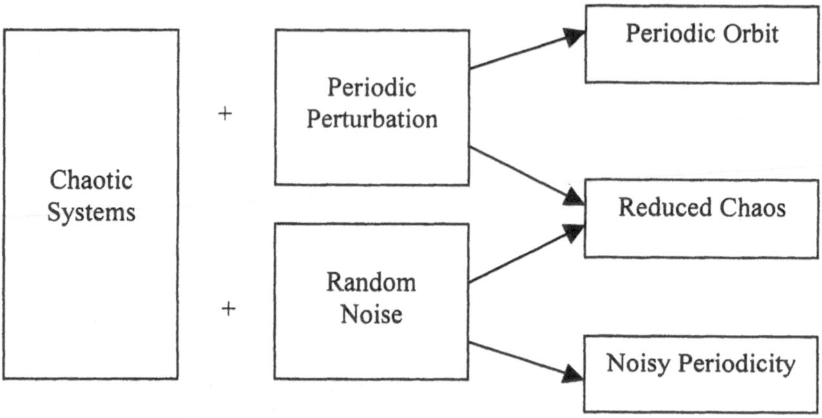

Figure 5.11 General idea of taming chaos.

5.2.3 Method Selection

Although the methods described in the previous sections have been developed mainly by physicists and mathematicians, generally most of them can be applied to control engineering systems.

In particular, the non-feedback methods can practically always be used. Their applications are straightforward and do not require special complicated controllers to be used. The main disadvantage of these methods is that the goal of controlling has to be determined by trial and error method.

The motivations for using feedback systems to control chaos are the following: feedback controllers are easy to implement, especially in electrical systems, they can perform the job automatically, and stabilize the overall control system efficiently. On the other hand, conventional feedback controllers are designed for non-chaotic systems. A chaotic system sensitivity to initial conditions may lead to the impression that in chaotic systems their sensitivity to small errors makes them very difficult. Such an impression may lead to the argument that once the control is initiated there is no need for further monitoring of the system's dynamics, nor feeding back this information in order to sustain the control. Indeed, it turns out that conventional feedback control of chaotic systems is generally difficult, but not impossible. Recently, Chen and Dong (1995) used a neural network approach for identification and control of chaotic systems. In many cases, a specially implemented feedback method can guarantee stabilization of the dynamic system. To summarize, the selection of the controlling method has to be based on the following:

(1) the goal of controlling (e.g. if the suppression of chaos is the main goal, then non-feedback methods can be applied in an easier way);
(2) the level of noise in the system (e.g. if the level of noise is large, then Pyragas's methods can be more effective than the OGY approach);
(3) the particular characteristics of the system. (Generally, in electrical systems one can try to use both feedback and non-feedback methods. In mechanical systems where the suppression of chaos is the main goal of controlling, non-feedback methods are recommended).

5.3 Summary

In this chapter, we have presented the main ideas underlying Dynamical Systems, Fractal theory, and Chaos theory, and we have only started to point out the many possible applications of these powerful mathematical theories. We have discussed in some detail the concepts of strange attractors, chaotic behavior and fractal dimension. The concept of the fractal dimension will be the basis of the method for time series analysis that will be used in Chapter 6, to achieve Mathematical Modelling of dynamic systems. Also, we have introduced several methods for controlling chaos that use some of the basic concepts of dynamical systems. These methods for controlling chaos can be used for real time control of dynamical systems, or can be used in combination with other computational techniques when the complexity of the problem requires more efficiency and accuracy. We will see in later chapter that many real world problems require hybrid control systems that combine several techniques to achieve the desired level of performance.

Chapter 6

Hybrid Intelligent Systems for Time Series Prediction

We describe in this chapter a new method for the estimation of the fractal dimension of a geometrical object using fuzzy logic techniques. The fractal dimension is a mathematical concept, which measures the geometrical complexity of an object. The algorithms for estimating the fractal dimension calculate a numerical value using as data a time series for the specific problem. This numerical (crisp) value gives an idea of the complexity of the geometrical object (or time series). However, there is an underlying uncertainty in the estimation of the fractal dimension because we use only a sample of points of the object, and also because the numerical algorithms for the fractal dimension are not completely accurate. For this reason, we have proposed a new definition of the fractal dimension that incorporates the concept of a fuzzy set. This new definition can be considered a weaker definition (but more realistic) of the fractal dimension, and we have named this the "fuzzy fractal dimension". We apply our fuzzy fractal approach to the problem of forecasting a particular time series, and compare our results to a neural network approach. The fuzzy fractal approach has some definite advantages over using neural networks, and we discuss these at the end of this chapter.

6.1 Problem of Time Series Prediction

Traditionally, we would assign a particular geometrical object a crisp value of the fractal dimension, and this numerical value was considered as a specific label for the object. However, this numerical value is difficult to use in classification or recognition applications because calculated values won't match these crisp values. We have experienced this problem when we used this idea for classifying bacteria with the fractal dimension (Castillo & Melin, 1994). We have found particular numerical labels for each of the bacteria, but when we try to use these values for recognizing specific bacteria in samples we have run into problems because of uncertainties. For this reason, we have proposed the following scheme for estimating the fuzzy fractal dimension of a set of geometrical objects. First, we calculate the standard fractal dimension of the objects, using the box counting algorithm with samples of points from the objects. Second, with the crisp values for the fractal dimensions of the objects build linguistic values for the dimensions, these will be the fuzzy fractal dimensions of the objects. Third, using these linguistic values of the fractal dimensions build a set of fuzzy rules that relate each object with each rule. This set of fuzzy if-then rules can be considered a classification scheme for the set of geometrical objects, and can be used for recognizing these objects because a particular value is mapped to an object. We can apply this method either for pattern recognition or for time series analysis as follows. First, we need to build the specific classification rules for the application using the fractal dimension. Then, we need to implement a method for sampling the object to obtain the data needed to calculate the crisp value of the fractal dimension. Finally, we use this crisp value as input in the set of fuzzy rules to obtain as output the specific classification for the object. For real image processing this can be used to recognize a particular object as needed for robotic applications (Castillo & Melin, 1998a). For time series analysis, this can be used for modeling and forecasting purposes. In any case, the generalization of the mathematical concept of the fractal dimension (Mandelbrot, 1987), to include now the ideas of fuzzy logic (Zadeh, 1975) is also important from the theoretical point of view because is only the initial point in the fuzzy generalization of Fractal Theory.

6.2 Fractal Dimension of an Object

Recently, considerable progress has been made in understanding the complexity of an object through the application of fractal concepts (Mandelbrot, 1987) and dynamic scaling theory. For example, financial time series show scaled properties suggesting a fractal structure (Castillo & Melin, 1999a). The fractal dimension of a geometrical object can be defined as follows:

$$d = \lim_{r \to 0} [\ln N(r)] / [\ln(1/r)] \qquad (6.1)$$

where N(r) is the number of boxes covering the object and r is the size of the box. An approximation to the fractal dimension can be obtained by counting the number of boxes covering the boundary of the object for different r sizes and then performing a logarithmic regression to obtain d (box counting algorithm). In Figure 6.1, we illustrate the box counting algorithm for a hypothetical curve C.

Counting the number of boxes for different sizes of r and performing a logarithmic linear regression, we can estimate the box dimension of a geometrical object with the following equation:

$$\ln N(r) = \ln\beta - d \ln r \qquad (6.2)$$

this algorithm is illustrated in Figure 6.2.

The fractal dimension can be used to characterize an arbitrary object. The reason for this is that the fractal dimension measures the geometrical complexity of objects. In this case, a time series can be classified by using the numeric value of the fractal dimension (d is between 1 and 2 because we are on the plane x y). The reasoning behind this classification scheme is that when the boundary is smooth the fractal dimension of the object will be close to one. On the other hand, when the boundary is rougher the fractal dimension will be close to a value of two.

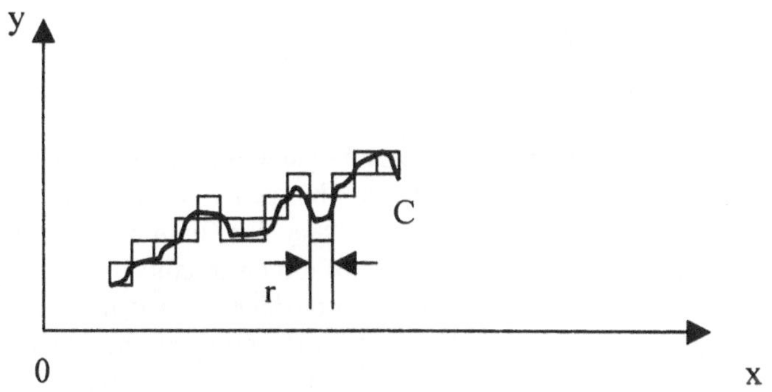

Figure 6.1 Box counting algorithm for a curve C.

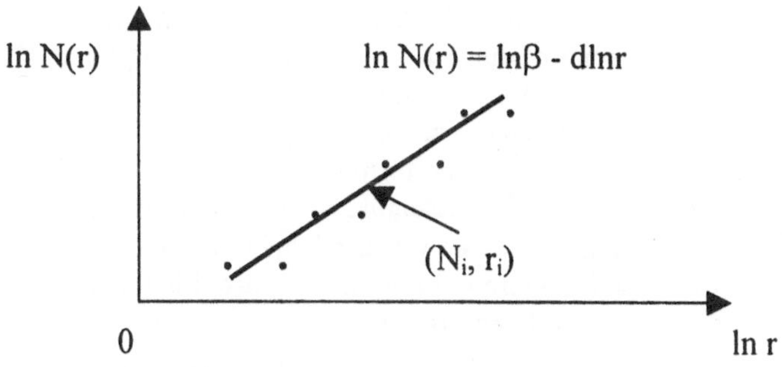

Figure 6.2 Logarithmic regression to find dimension.

6.3 Fuzzy Logic for Object Classification

We can use a fuzzy rule base as a classification scheme if we can make a suitable partition of the input space such that we are able to distinguish different geometrical objects by their characteristics. We will consider geometrical objects in the plane for simplicity. We can now use fuzzy clustering techniques (Yager & Filev, 1994) to cluster the data, and then build a fuzzy rule base that will actually be a classification scheme for the particular application.

We will consider that we have n objects O1, O2, ..., On, and that we are able to apply fuzzy clustering techniques to obtain n pairs (Xi, Yi) i=1,...,n, which are the respective centers of the n clusters. Then a fuzzy rule base can be constructed as follows:

$$\begin{array}{l} \textbf{If } X \text{ is } x_1 \text{ and } Y \text{ is } y_1 \textbf{ then } \text{Object is } O_1 \\ \textbf{If } X \text{ is } x_2 \text{ and } Y \text{ is } y_2 \textbf{ then } \text{Object is } O_2 \\ \ldots \qquad \ldots \qquad \ldots \\ \textbf{If } X \text{ is } x_n \text{ and } Y \text{ is } y_n \textbf{ then } \text{Object is } O_n \end{array} \qquad (6.3)$$

These rules can be used for pattern classification or time series analysis because in both cases the data has the same general structure. For applications of higher dimensionality this approach can be generalized in a straightforward manner, but of course the problem is that the number of rules increases dramatically (which is known as the curse of dimensionality).

To illustrate these ideas we will consider a particular application. Lets consider the problem of forecasting the time series of the exchange rate US dollar/MX peso. We used the time series of average weekly rates for 36 weeks to find the fuzzy model as explained above. We then used the fuzzy model to predict future values of the exchange rate and compare these to the actual values to validate this approach.

We show in Figure 6.3 the time series of exchange rates for 36 weeks of US dollar/MX peso from August 1999 to April 2000. We can notice from this figure the cyclical behavior of the time series over this short period of time.

We used the Fuzzy Logic Toolbox of MATLAB for fuzzy clustering of this data, and then for implementing the fuzzy rule base using the recognized clusters. In this case, five rules of the form shown in Equation 6.3 were used. We show in Figure 6.4 the general architecture of the fuzzy system. In this case, the Mamdani fuzzy reasoning procedure was used due to its simplicity.

We also show in Figure 6.5 how we can use this fuzzy system to predict a particular value of the exchange rate base on the actual value of the exchange rate and the value of future time. We validated these results against the real values that occurred during the following four weeks.

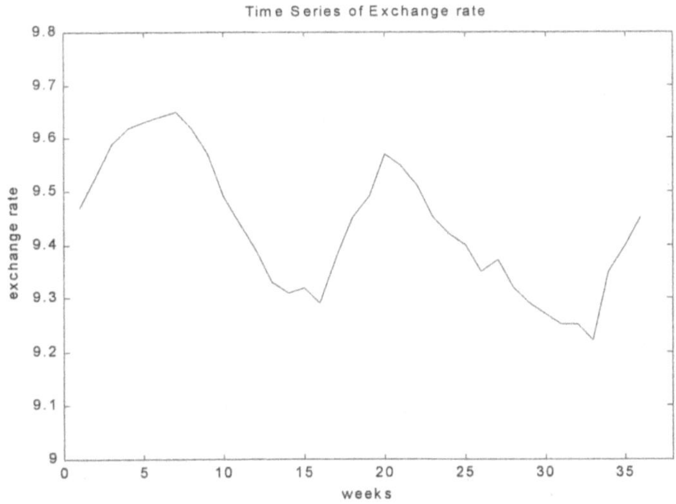

Figure 6.3 Time series of exchange rates US/Mexico.

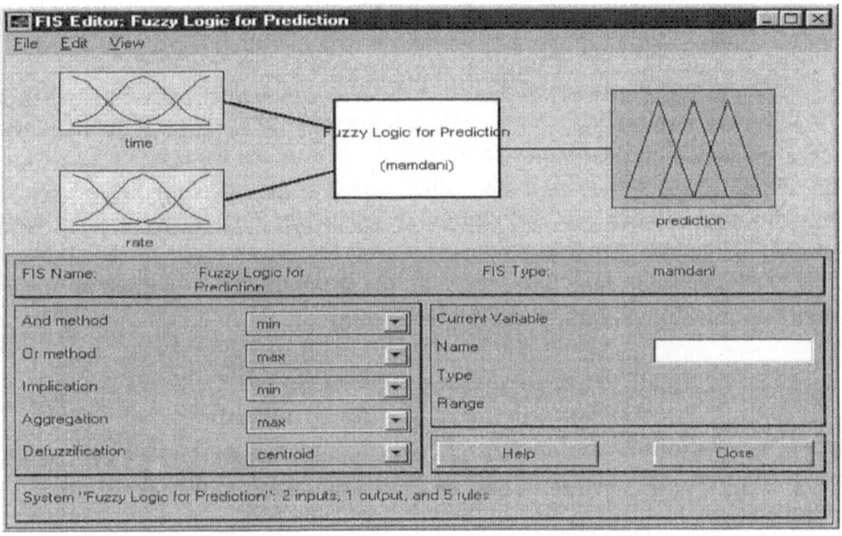

Figure 6.4 General Architecture of the fuzzy system.

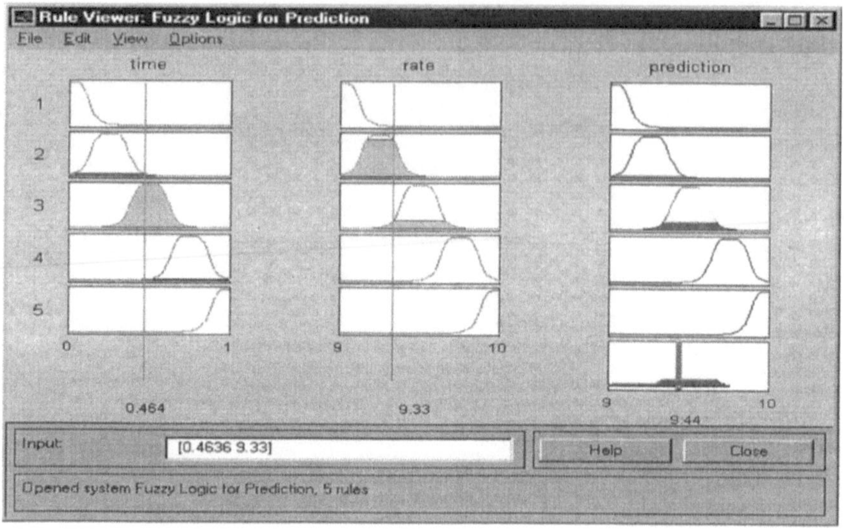

Figure 6.5 Fuzzy Prediction of the exchange rate.

6.4 Fuzzy Estimation of the Fractal Dimension

The fractal dimension of a geometrical object is a crisp numerical value measuring the geometrical complexity of the object. However, in practice it is difficult to assign a unique numerical value to an object. It is more appropriate to assign a range of numerical values in which there exists a membership degree for this object. For this reason, we will assign to an object O a fuzzy set μ_o, which measures the membership degree for that object.

Lets consider that the object O is in the plane x y, then a suitable membership function is a generalized bell function:

$$\mu_o = 1 / [\, 1 + |\, (x-c) / a\, |^{2b} \,] \tag{6.4}$$

where a, b and c are the parameters of the membership function. Of course other types of membership functions could be used depending on the characteristics of the application.

By using the concept of a fuzzy set we are in fact generalizing the mathematical concept of the fractal dimension. In fact, our definition of the fuzzy fractal dimension is as follows.

Definition 6.1: Let O be an arbitrary geometrical object in the plane x y. Then the fuzzy fractal dimension is the pair: (d_o, μ_o)
where d_o is the numerical value of the fractal dimension calculated by the box counting algorithm, and μ_o is the membership function for the object.

With this new definition we can account for the uncertainty in the estimation of the fractal dimension of an object. Also, this new definition enables easier pattern recognition for objects, because it is not necessary to match an exact numerical value to recognize a particular object.

6.5 Fuzzy Fractal Approach for Time Series Analysis and Prediction

Let us consider now the problem of time series analysis and prediction. Let y_1, y_2, ..., y_n be an arbitrary time series. If we want to be able to forecast this time series, we need to analyze the data and extract the trends and periodicities of the series. Assuming that the time series can be clustered into n objects O_1, O_2, ..., O_n as shown in Figure 6.6, then we can build a fuzzy rule base as in Section 6.3 of this chapter. However, we now also want to consider the geometrical complexity of the objects O_1, O_2, ..., O_n as measured by their fractal dimensions d_{o1}, d_{o2}, ..., d_{on} respectively. Then a fuzzy rule base for time series prediction can be expressed as follows.

If dim is d_{o1} and pos is x_1 **then** prediction is O_1
If dim is d_{o2} and pos is x_2 **then** prediction is O_2
... ... (6.5)
If dim is d_{on} and pos is x_n **then** prediction is O_n

In this case, we need to define membership functions for the fractal dimension, position, and for the geometrical objects. This fuzzy rule base can be used with the Mamdani reasoning method, and center of gravity as defuzzification method. However, it is also possible to use a Sugeno type fuzzy system in which the consequents can be linear functions. This is illustrated in Equation (6.6).

If dim is d_{o1} and pos is x_1 **then** $y=a_1x_1+b_1d_{o1}+c_1$
If dim is d_{o2} and pos is x_2 **then** $y=a_2x_2+b_2d_{o2}+c_2$
... ... (6.6)
If dim is d_{on} and pos is x_n **then** $y=a_nx_n+b_nd_{on}+c_n$

In this case, we can use a neuro-fuzzy approach for adapting the parameters of the fuzzy system using real data of the problem. We can use, for example, an ANFIS approach (Jang, Sun & Mizutani, 1997) to learn from real data the best values for the coefficients of the linear functions and for the membership functions.

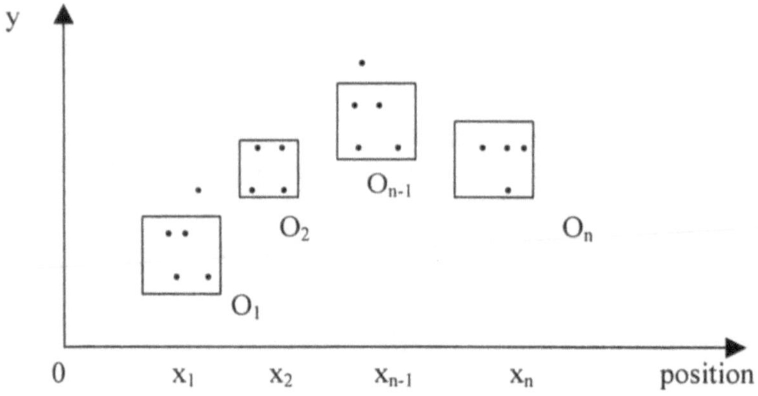

Figure 6.6 Fuzzy clustering of the time series.

We show in Figure 6.7 an implementation of the Mamdani type fuzzy system in the Fuzzy Logic Toolbox of MATLAB for the time series of exchange rates of US dollars/MX pesos. In this figure, we can see the non-linear surface for the fuzzy inference system of prediction.

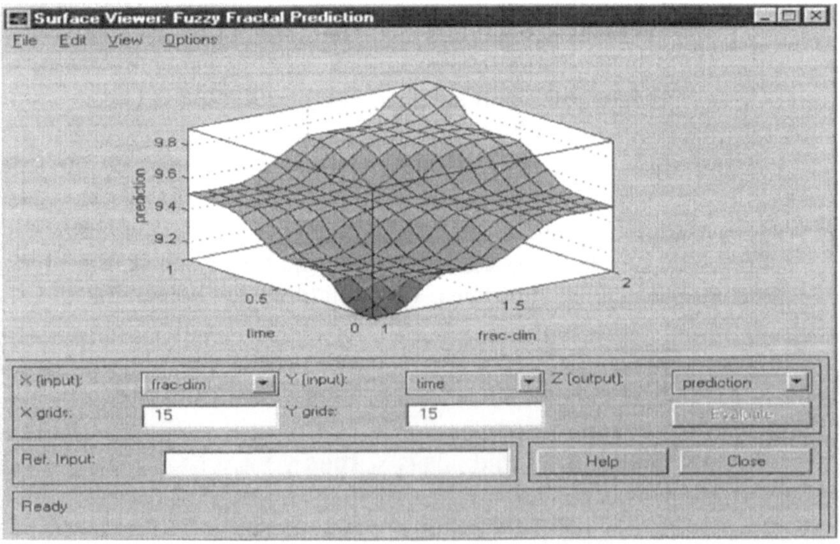

Figure 6.7 Non-Linear Surface for time series.

We also show in Figure 6.8 the implementation of the Sugeno type fuzzy system for the same time series. In this figure, we can see the non-linear surface for the Sugeno fuzzy system for time series prediction. This surface represents the fuzzy model for the problem of predicting the exchange rate of the US dollar/MX peso.

Finally, we show in Figure 6.9 the fuzzy reasoning for prediction for particular values of the fractal dimension and time.

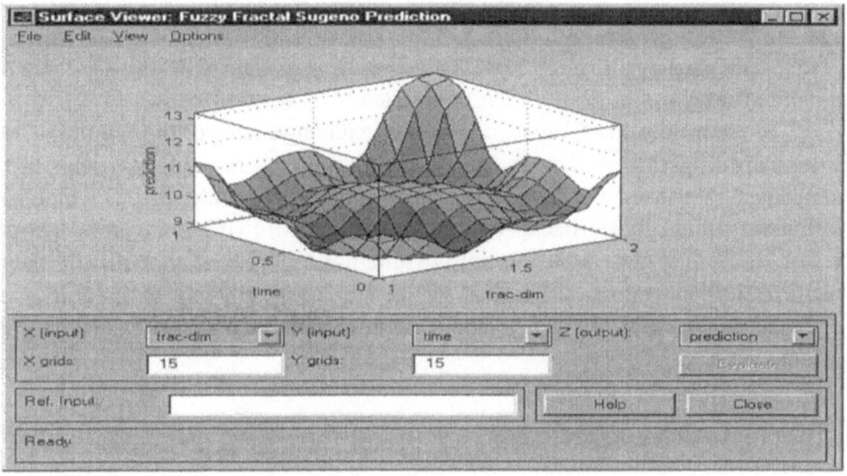

Figure 6.8 Surface for Sugeno type fuzzy system.

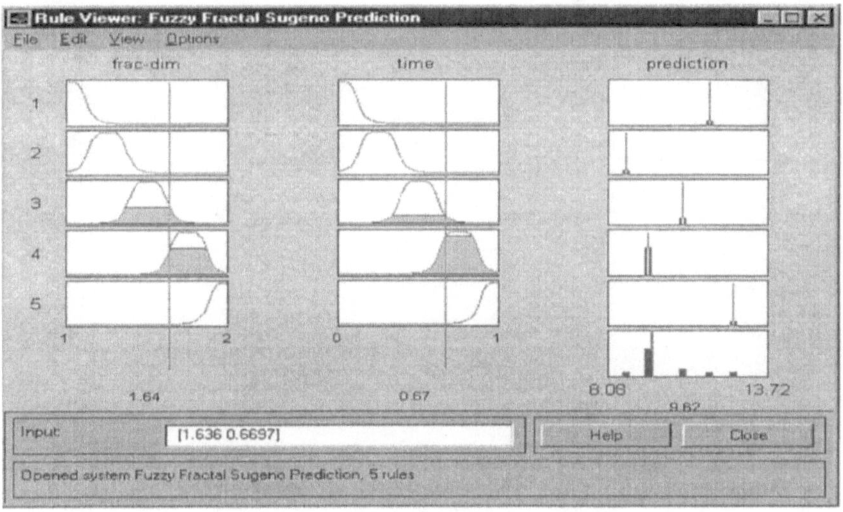

Figure 6.9 Fuzzy reasoning for time series prediction.

6.6 Neural Network Approach for Time Series Prediction

We can also use neural networks for time series analysis and prediction. In this case, a neural network is given a time series as training data, and then the forecasting ability of the network is measured with new data. We use a feedforward neural network with the Levenberg-Marquardt learning algorithm for time series prediction, to compare the results of this approach with our fuzzy fractal methodology. We will first present our results with the neural network approach, and then we discuss how these results compare with the ones that we obtained before with the fuzzy fractal approach.

We consider the same time series of exchange rates of the US dollar/MX peso to train a feedforward neural network of three layers, with 15 nodes in the hidden layer. We used the Levenberg-Marquardt training algorithm, which is a modification of the basic backpropagation method allowing a variable learning rate. We show in Figure 6.10 the initial function approximation with the neural network described before. Of course, at the beginning the approximation is not good because the network is initialized with random weights by the computer program.

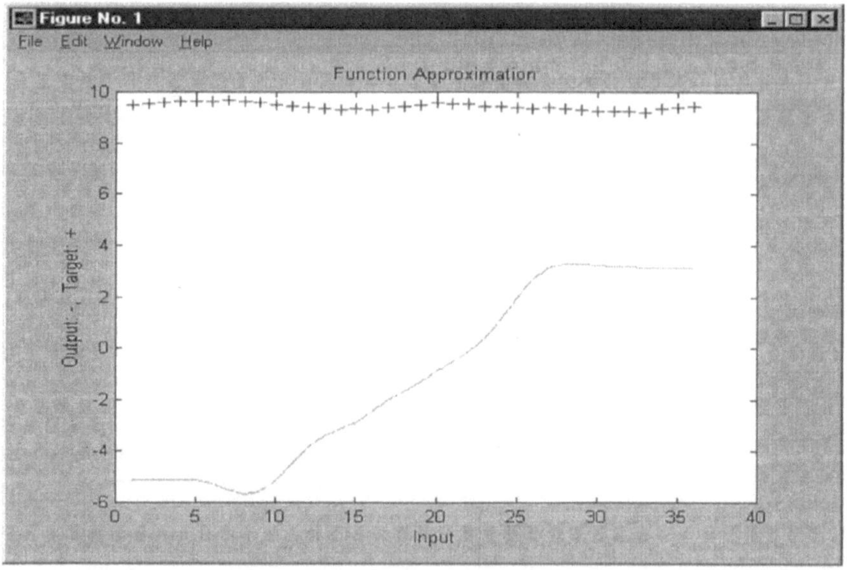

Figure 6.10 Initial function approximation with the neural network.

We show in Figure 6.11 the results of training the neural network for 420 epochs. The sum of squared errors after training was about 0.0002, which satisfied our goal error in the computer program. We can see in this figure, that the network follows very closely the training data. Also, the results of the neural network were compared to the real values of the exchange rate for the following weeks after training, and forecasts were always within a 5% error of the real values.

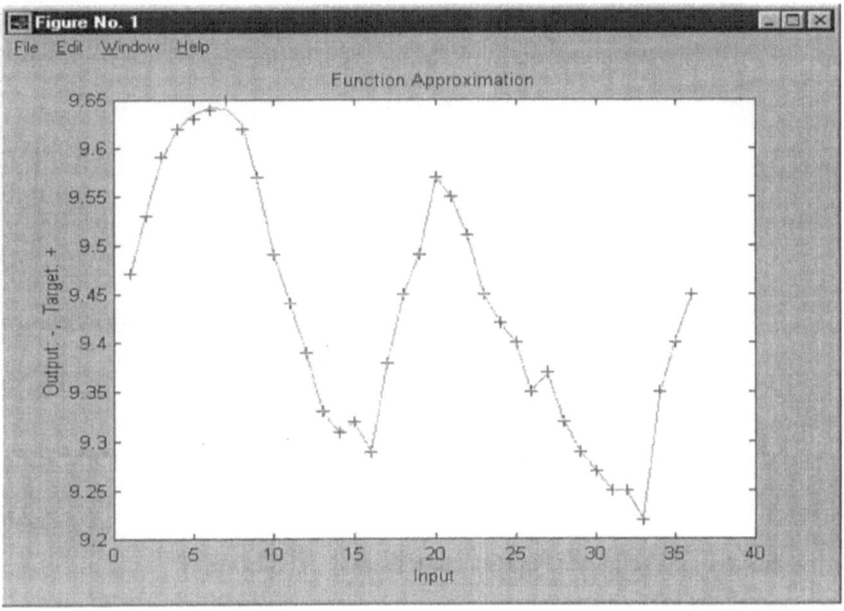

Figure 6.11 Final function approximation of the neural network.

Finally, we show in Figure 6.12 the results of plotting the sum of squared errors against the number of epochs. We can see from this figure how the neural network learns very rapidly the pattern of the time series.

The results of the neural network can be considered very good if we validate in the short term, as mentioned before. The problem of these type of results is that when the time series starts to behave more erratically then the network loses its forecast ability. Here is when a fuzzy fractal approach becomes more powerful because it has more flexibility in adapting to different types of situations. Our fuzzy fractal approach also has short term forecast ability (within 5% of real values), but at the same time is able to give good results in the long term. This is true because the fuzzy rule base contains a kind of knowledge about the time series.

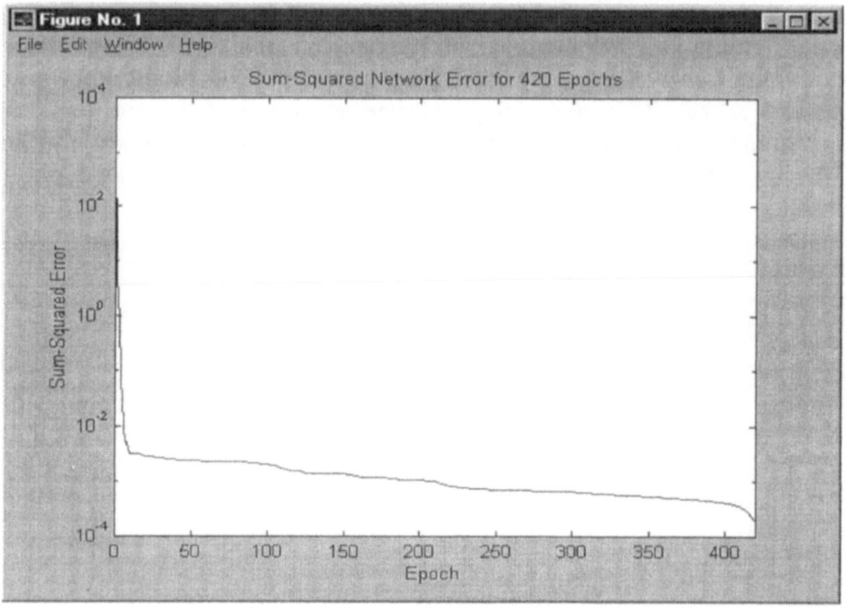

Figure 6.12 Sum of squared errors against the number of epochs.

6.7 Fuzzy Fractal Approach for Pattern Recognition

We can also use the above ideas for pattern recognition in image processing applications. The method is very similar to the one for time series prediction, the only difference is that the data is not directly related to time. For pattern recognition only real geometrical variables are used. In this case, we also consider n objects O_1, O_2, ..., On with n corresponding cluster centers (x_i, y_i), i=1,...,n. Then the fuzzy rule base can be stated as follows.

If dimension is d_{o1} **then** Object is O_1
If dimension is d_{o2} **then** Object is O_2
... ... (6.7)
If dimension is d_{o2} **then** Object is O_n

To completely define this fuzzy system for pattern recognition, we will need to define the membership functions for the fractal dimensions and the objects. The method for calculating the fractal dimension is the same as before and for fuzzy reasoning we can use Mamdani or Sugeno type.

6.8 Summary

We have presented in this chapter a new approach for time series analysis and pattern recognition combining fuzzy logic and fractal theory. With the new approach, we can always build a set of fuzzy rules using the fractal dimension of the objects to solve the problem of forecasting or recognition. We have shown very good results in predicting the exchange rate of the US dollar/MX peso with this new approach. Our new method can also be applied to similar problems of prediction. We compared the results of our approach with the ones given by a neural network, and we discuss the advantages of using a fuzzy rule base instead of a network. We have to mention here that the problem of time series analysis and prediction is very important in the real world, because we are always in need of forecasting the future behavior of a dynamical system to be able to control it or for decision making.

Chapter 7

Modelling Complex Dynamical Systems with a Fuzzy Inference System for Differential Equations

We describe in this chapter a new method for modelling complex dynamical systems based on the use of a new fuzzy inference system for multiple differential equations. It is well known that formulating a unique and sufficiently accurate mathematical model for a complex dynamical system (over a whole region of discourse) may be very difficult or even impossible in some cases (Castillo and Melin, 1996). For this reason, it may be more efficient to formulate a set of mathematical models that approximate the local behavior of the dynamical system for different parameter regions. We can then formulate a set of fuzzy if-then rules relating these regions to their corresponding mathematical models. We can assume, without loss of generality that the models can be expressed as non-linear differential equations (Castillo and Melin, 1997). We have developed a fuzzy inference system that enables fuzzy reasoning with multiple differential equations. The new fuzzy system can be considered as a generalization of Sugeno's inference procedure, in which we are now using differential equations as consequents of the fuzzy rules instead of simple polynomials like in Sugeno's original method. We illustrate our new method for modelling with two cases: robotic dynamic systems, and aircraft systems. These two applications are complex enough to illustrate the power of our method for modelling.

7.1 The Problem of Modelling Complex Dynamical Systems

The classical Sugeno fuzzy inference system (also known as the TSK fuzzy model) was proposed by Takagi, Sugeno, and Kang (1985) in an effort to develop a systematic approach to generating fuzzy rules from a given input-output data set. A typical fuzzy rule in a Sugeno fuzzy model has the form

$$\textbf{If } x \text{ is } A \text{ and } y \text{ is } B \textbf{ then } z = f(x,y) \qquad (7.1)$$

where A and B are fuzzy sets in the antecedent, while z = f(x,y) is a crisp function in the consequent. Usually f(x,y) is a polynomial in the input variables x and y, but it can be any function as long at it can appropriately describe the output of the model within the fuzzy region specified by the antecedent of the rule. When f(x,y) is a first order polynomial, the resulting fuzzy inference system is called a "first-order Sugeno fuzzy model", which was originally proposed in (Sugeno & Kang, (1988). When f is a constant, we then have a "zero-order Sugeno fuzzy model", which can be viewed as a special case of the Mamdani inference system. The overall output of the Sugeno fuzzy model is obtained via a weighted average operator, thus avoiding the time-consuming process of defuzzification required by a Mamdani model (Mamdani & Assilian, 1975).

Our new fuzzy inference system uses differential equations as consequents in the rules, instead of simple polynomials. The new fuzzy inference system can be considered a generalization of Sugeno's original inference system, because we are now modelling a particular problem by using the appropriate differential equation for each region of the domain. A typical rule in this case has the form

$$\textbf{If } x \text{ is } A \text{ and } y \text{ is } B \textbf{ then } dz/dt = f(x,y) \qquad (7.2)$$

where A and B are fuzzy sets in the antecedent, while dz/dt = f(x,y) is crisp differential equation in the consequent. Usually f(x,y) is a non-linear function of the input variables x and y, and this means that we have a non-linear differential equation in the consequent. We have to note here that this new fuzzy inference system reduces to the standard Sugeno system only when the differential equations have closed form solutions in the form of polynomials. However, the solutions to the differential equations can be more complicated analytic functions or in most cases the solutions are so complex that can only be approximated by numerical methods. The advantage of this generalization of Sugeno's original method is that, in general, we can represent more complicated dynamic behaviors and also because of this fact, the number of rules needed to represent a given dynamical system is smaller.

7.2 Modelling Complex Dynamical Systems with the new Fuzzy Inference System

For a real-world dynamical system it may be necessary to consider a set of mathematical models to represent adequately all of the possible dynamic behaviors of the system (Castillo & Melin, 1998b). In this case, we need a fuzzy decision procedure to select the appropriate model to use according to the value of a selection parameter vector α. To implement this decision procedure, we need a fuzzy inference system that can use differential equations as consequents. For this purpose, we have developed a new fuzzy inference system that can be considered as a generalization of Sugeno's inference system (Sugeno & Kang, 1988), in which we are now using differential equations as consequents of the fuzzy rules, instead of simple polynomials like in the original Sugeno's method. Using this method, a fuzzy model for a general dynamical system can be expressed as follows:

$$
\begin{array}{l}
\textbf{IF } \alpha_1 \text{ is } A_{11} \textbf{ AND } \alpha_2 \text{ is } A_{12} \ldots \textbf{ AND } \alpha_m \text{ is } A_{1m} \\
\qquad\qquad\qquad\qquad \textbf{THEN } dy/dt = f_1(y, \alpha) \\
\textbf{IF } \alpha_1 \text{ is } A_{21} \textbf{ AND } \alpha_2 \text{ is } A_{22} \ldots \textbf{ AND } \alpha_m \text{ is } A_{2m} \\
\qquad\qquad\qquad\qquad \textbf{THEN } dy/dt = f_2(y, \alpha) \\
\vdots \qquad\qquad\qquad\qquad\qquad\qquad \vdots \\
\textbf{IF } \alpha_1 \text{ is } A_{n1} \textbf{ AND } \alpha_2 \text{ is } A_{n2} \ldots \textbf{ AND } \alpha_m \text{ is } A_{nm} \\
\qquad\qquad\qquad\qquad \textbf{THEN } dy/dt = f_n(y, \alpha)
\end{array}
\tag{7.3}
$$

where A_{ij} is the linguistic value of α_j for rule i-th, $\alpha \in R^m$ and is defined by $\alpha = [\alpha_1, \ldots, \alpha_m]$, and $y \in R^p$ is the output obtained by the numerical solution of the corresponding differential equation. Of course, it is assumed that each differential equation in (7.3) locally approximates the real dynamical system over a neighborhood (or region) of R^m.

The numerical solution of the differential equations can be achieved by the standard Runge-Kutta type method (Nakamura, 1997):

$$
\begin{aligned}
y_{n+1} &= RK\,(y_n) + 1/2(k_1 + k_2) \\
k_1 &= hf(y_n, t_n) \\
k_2 &= hf(y_n + k_1, t_{n+1})
\end{aligned}
\tag{7.4}
$$

where h is the step size of the method and RK can be considered as the Runge-Kutta operator that transforms numerical solutions from time n to time n+1. Numerical solutions are then aggregated by weighted average with weights obtained by the minimum of the firing strengths of the inputs:

$$
y = \frac{w_1 y_1 + w_2 y_2 + \ldots + w_n y_n}{w_1 + w_2 + \ldots w_n} \tag{7.5}
$$

where:

$$y_1 = RK(f_1(y, \alpha)),\ y_2 = RK(f_2(y, \alpha)),\ \ldots,\ y_n = RK(f_n(y, \alpha)).$$

The new fuzzy inference system for differential equations can be illustrated as in Figure 7.1, where a complex dynamical system is modeled by using four different mathematical models (M_1, M_2, M_3 and M_4). The decision scheme can be expressed as a single-input fuzzy model as follows:

$$\begin{array}{llll} \textbf{IF} & \alpha \text{ is small} & \textbf{THEN} & dy/dt = f_1(y, \alpha) \\ \textbf{IF} & \alpha \text{ is regular} & \textbf{THEN} & dy/dt = f_2(y, \alpha) \\ \textbf{IF} & \alpha \text{ is medium} & \textbf{THEN} & dy/dt = f_3(y, \alpha) \\ \textbf{IF} & \alpha \text{ is large} & \textbf{THEN} & dy/dt = f_4(y, \alpha) \end{array} \qquad (7.6)$$

where the output y is obtained by the numerical solution of the corresponding differential equation.

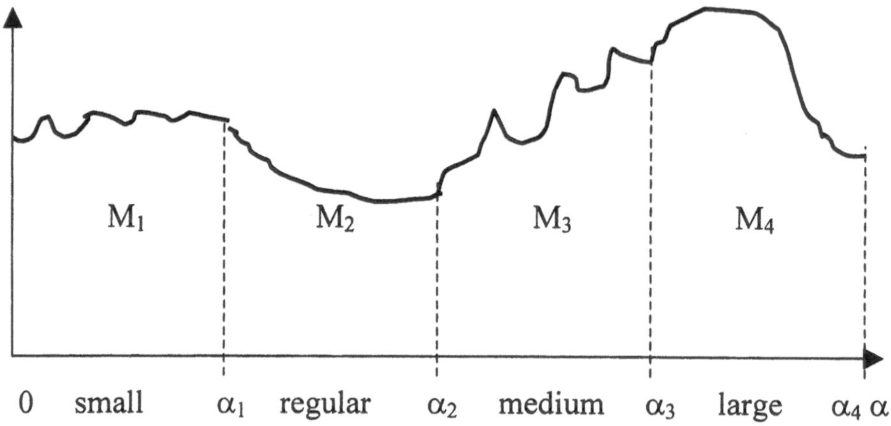

Figure 7.1 Modelling a complex dynamical system with the new fuzzy inference system.

7.3 Modelling Robotic Dynamic Systems with the New Fuzzy Inference System

We describe in this section the application of the new fuzzy inference method for differential equations to the case of robotic dynamic systems. We first review the general mathematical models for robotic systems. We then show how to use the new fuzzy reasoning methodology to the problem of modelling complex robotic dynamic systems. Finally, we show results of the simulations of the fuzzy system for modelling. These results closely resemble the real dynamic behaviors of these robotic systems.

7.3.1 Mathematical Modelling of Robotic Systems

In the last several years, many papers have been published, rendering an important contribution to the development of computer methods for the mathematical modelling of robotic systems. The modelling methods may be classified with respect to the laws of mechanics on the basis of which motion equations are formed. One may distinguish methods based on Lagrange, Newton-Euler, D'Alembert, and other formalisms for dynamic modelling of interconnected multibody systems. The dynamic model of the robot consists of the model of the mechanical part of the robot (mechanism) and the model of the actuators that are driving the robot joints. The model of the mechanical part of the robot is usually assumed, see Vukobratovic' (1989), in the following form:

$$P = H(q)\, q'' + h(q, q') \tag{7.7}$$

where: $P = n \times 1$ vector of driving torques in the joints, $H = n \times n$ inertia matrix of the mechanism, $h = n \times 1$ vector of centrifugal, Coriolis, and gravity moments (forces) around the axes of the joints.

Various types of actuators are applied to drive robots: dc motors, ac motors, hydraulic actuators, pneumatic actuators and so on. The models of actuators are in general non-linear, but for the dc motors (which are still most often applied for industrial robots) a linear state model may be used:

$$(X^i)' = A^i X^i + b^i u^i + f^i P_i, \qquad i = 1, 2, \ldots, n \tag{7.8}$$

where: $X^i = (q_i, q_i', i_{Ri})^T = 3 \times 1$ state vector of ith actuator model, i_{Ri} = rotor current of ith dc motor, u^i = scalar input to ith actuator, P_i = driving torque (load) in ith joint, $A^i = 3 \times 3$ matrix, $b^i, f^i = 3 \times 1$ vectors.

The connection between the models (7.7) and (7.8) (through the state coordinates q_i, q_i', and driving torques P_i) are evident. Certain constraints upon the actuators input u amplitude as well as on the allowable driving torques should be also added to these models.

7.3.2 Fuzzy Modelling of Robotic Dynamic Systems

We will consider, in this section, the case of modelling robotic manipulators to illustrate the application of the method for fuzzy modelling complex dynamical systems. The general mathematical model for this kind of robotic system is the following (Zomaya, 1992):

$$M(q)q'' + V(q, q'))q' + G(q) + F_d q' = \tau \qquad (7.9)$$

where $q \in R^n$ denotes the link position, $M(q) \in R^{nxn}$ is the inertia matrix, $V(q,q') \in R^{nxn}$ is the centripetal-Coriolis matrix, $G(q) \in R^n$ represents the gravity vector, $F_d \in R^{nxn}$ is a diagonal matrix representing the friction term, and τ is the input torque applied to the links.

For the simplest case of a one-link robot arm (Yamamoto & Yun, 1997), we have the scalar equation:

$$M_q q'' + F_d q' + G(q) = \tau \qquad (7.10)$$

If G(q) is a linear function (G = Nq), then we have the "linear oscillator" model:

$$q'' + aq' + bq = c$$

where $a = F_d/M_q$, $b = N/M_q$ and $c = \tau/M_q$. This is the simplest mathematical model for a one-link robot arm. More realistic models can be obtained for more complicated functions G(q). For example, if $G(q) = Nq^2$, then we obtain the "quadratic oscillator" model:

$$q'' + aq' + bq^2 = c \qquad (7.11)$$

where a, b and c are defined as above.

A more interesting model is obtained if we define G(q) = Nsinq. In this case, the mathematical model is

$$q'' + aq' + b\sin q = c \qquad (7.12)$$

where a, b and c are the same as above. This is the so-called "sinusoidally forced oscillator". More complicated models for a one-link robot arm can be defined similarly.

For the case of a two-link robot arm, we can have two simultaneous differential equations as follows:

$$q''_1 + a_1 q'_1 + b_1 q^2_2 = c_1 \qquad (7.13)$$

$$q''_2 + a_2 q'_2 + b_2 q^2_1 = c_2$$

which is called the "coupled quadratic oscillators" model. In Equation (7.13) a_1, b_1, a_2, b_2, c_1 and c_2 are defined similarly as in the previous models. We can also have the "coupled cubic oscillators" model:

$$q''_1 + a_1 q'_1 + b_1 q^3{}_2 = c_1 \qquad (7.14)$$
$$q''_2 + a_2 q'_2 + b_2 q^3{}_1 = c_2$$

We can also have the "coupled forced quadratic oscillators" model:

$$q''_1 + a_1 q'_1 + b_1 q^2{}_1 = c_1 \sin q_2 \qquad (7.15)$$
$$q''_2 + a_2 q'_2 + b_2 q^2{}_2 = c_2 \sin q_2$$

which is a system of two coupled second-order non-linear differential equations. More complicated models for a two-link robot arm can be defined similarly.

We use a fuzzy rule base for model selection for the case of robotic manipulators. We presented before mathematical models that can be used to model the dynamic behavior of robotic manipulators. Lets call M_1 the mathematical model given by Equation (7.11), M_2 the mathematical model given by Equation (7.12), M_3 the model given by Equation (7.13), and M_4 the model given by Equation (7.14). Then we can establish a fuzzy rule base for these models as explained in Section 2 of this chapter. We will assume here without loss of generality that the selection parameters are the fractal dimension of a time series of measured values of the relevant variables in the problem (angle, angular velocity) and the number of links of the manipulator. Also, we are assuming that only four models are needed to model completely the robotic system. Then, we can define a set of four fuzzy if-then rules that basically relate the fuzzy values of the selection parameters with the corresponding mathematical model. We show in Table 7.1 this set of fuzzy rules for model selection for the case of manipulators of one and two links.

We also need to define the membership functions for the fuzzy values in Table 7.1. The membership functions for the models should give us the degree of belief that a particular mathematical model is the correct one for the specific values of the selection parameters. We have to note here that for using a fuzzy rule base (like the one described in Table 7.1) with mathematical models, we need to use our new fuzzy inference system for multiple differential equations (described in Section 2 of this chapter).

Table 7.1 Fuzzy rule base for model selection of robotic systems

IF		THEN
Fractal dimension	Number of links	Model
low	one	M_1
high	one	M_2
low	two	M_3
high	two	M_4

7. 3. 3 Experimental Results

To give an idea of the performance of our new method for modelling complex dynamical systems, we show below simulation results for several types of robotic systems. First, we show in Figure 7.2 the membership functions for the fractal dimension variable. The membership functions were defined in the membership function editor of the fuzzy logic toolbox of MATLAB.

We show in Figure 7.3 the non-linear surface for the problem of modelling robotic manipulators using as input variables: the fractal dimension and number of links. The three dimensional surface represents the non-linear fuzzy model for the problem.

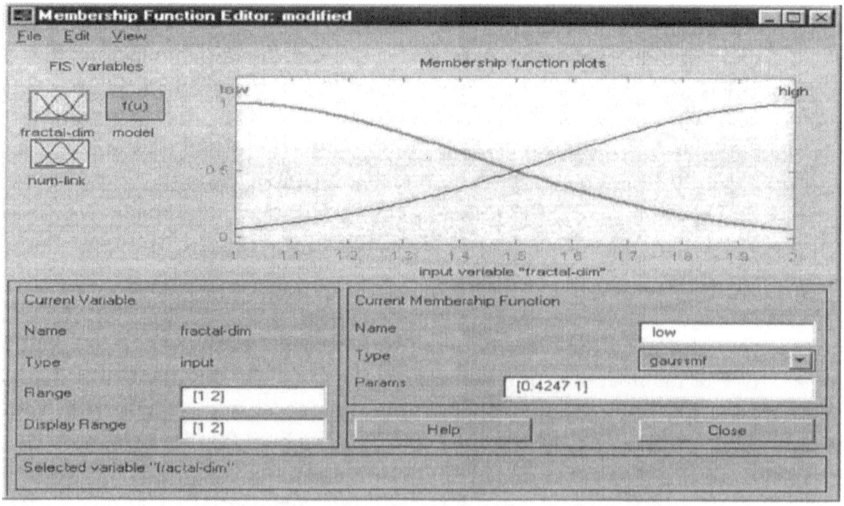

Figure 7.2 Membership function plots for the linguistic values of the fractal dimension variable.

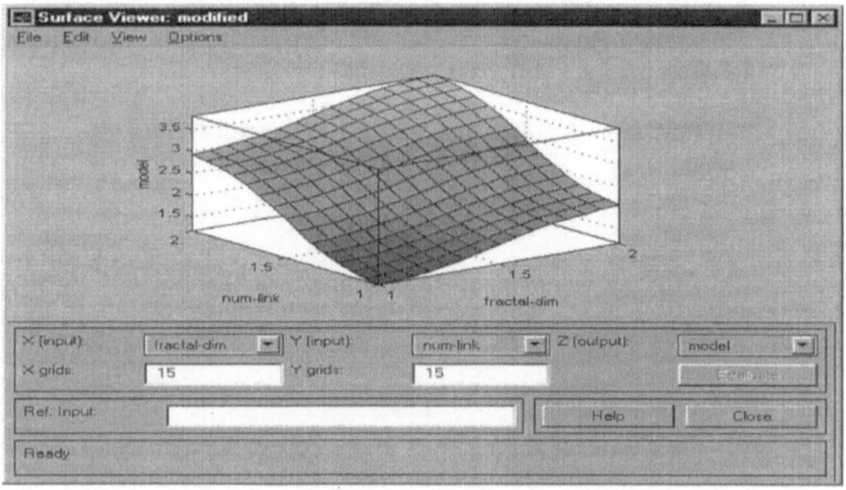

Figure 7.3 Non-linear surface for modelling robotic manipulators.

We show in Figure 7.4 the reasoning procedure for modelling robotic dynamic systems when specific values for the fractal dimension and number of links are given. In this figure we can see how the final output (combined) model is obtained with the new fuzzy inference system. The results correspond to the values for the domain of application.

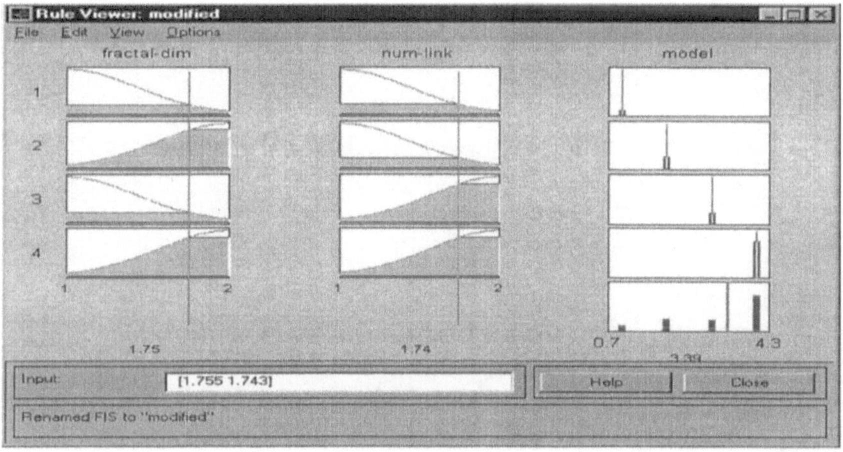

Figure 7.4 Reasoning procedure for specific values of the fractal dimension and number of links.

We show in Figure 7.5 the general structure of the fuzzy intelligent system for modelling robotic dynamic systems, developed with the fuzzy logic toolbox of the MATLAB programming language.

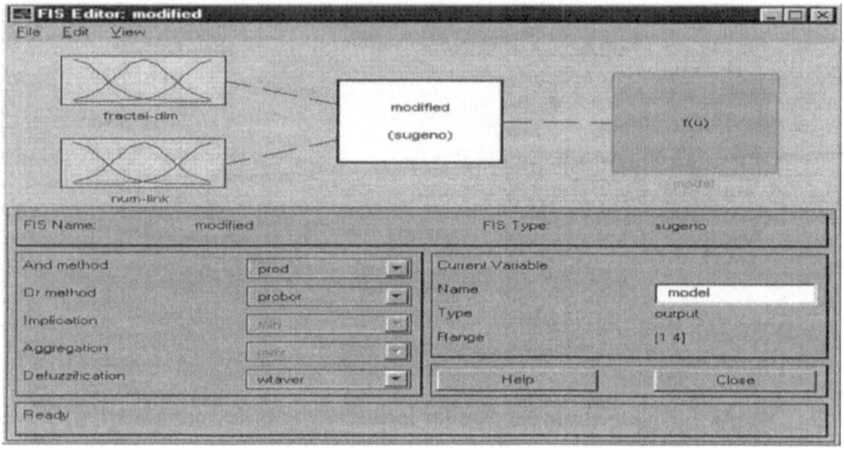

Figure 7.5 General structure of the fuzzy intelligent system with the new fuzzy inference system.

We show simulation results for a robotic system obtained using our new method for modelling dynamical systems. In Figure 7.6 we show the simulation results for a two-link robotic dynamic system with a mathematical model given by Equation (7.13).

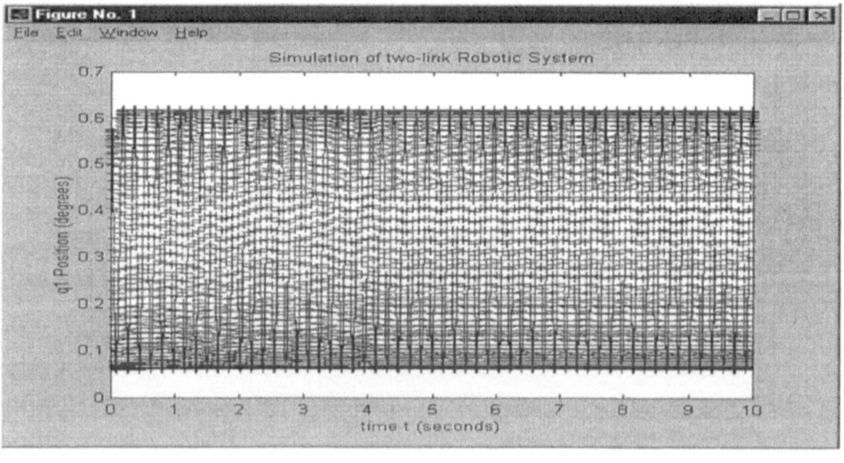

Figure 7.6 Simulation of a two-link robot arm showing chaotic behavior for position q_1.

The solution shown in Figure 7.6 is what is known as a "chaotic solution" because the orbit is oscillating (in an unstable manner) between an infinite number of periodic points. As a consequence of this fact the behavior identification in this case is of a "chaotic solution".

In Figure 7.7 we can see how both q_1 and q_2 tend to a "strange attractor", which is one of the distinguishing signs of "chaotic" behavior (Rasband, 1990). Of course, in robotic applications this behavior has to be avoided because it will cause physical damage to the robotic system. This is why it is important to identify when this behavior can occur in advance to avoid critical situations.

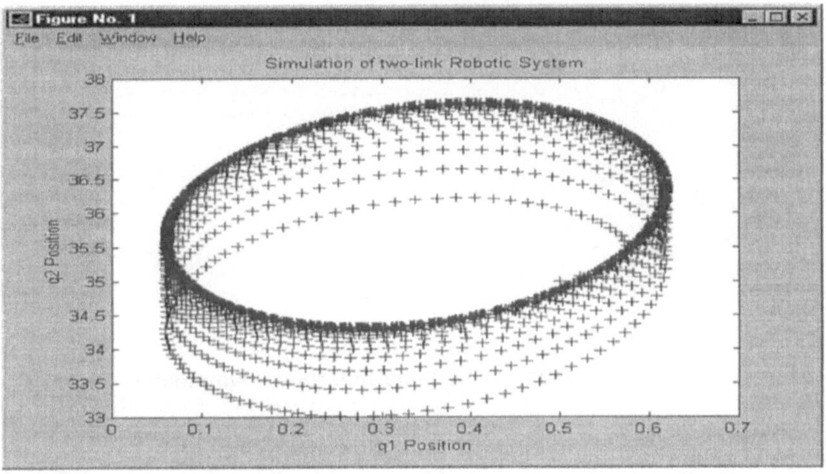

Figure 7.7 Simulation of a two-link robot arm showing chaotic behavior for positions q_1 and q_2.

Finally, we show in Figure 7.8 simulation results for a one-link robotic manipulator, were the dynamic behavior is a cycle of period 8. This dynamic behavior shows the existence of a cascade of period doubling bifurcations (and eventually chaos) for a single-link robotic system.

We have developed a new fuzzy inference system for modelling complex dynamical systems using multiple differential equations. The new fuzzy inference system can be considered a generalization of the classical Sugeno fuzzy model. The use of differential equations as consequents instead of polynomials, gives to our new fuzzy modelling procedure more approximating power and as a result of this fact, a smaller number of fuzzy rules are needed to model a given problem. We have illustrated in this paper our new method for modelling with the complex case of robotic dynamic systems. Our new fuzzy inference system allows the change of mathematical models according to the changing conditions of the robotic system and its environment. Adequate modelling of complex dynamic systems (like the robotic system) enables adaptive model-based control, which is a very important problem in real-world applications.

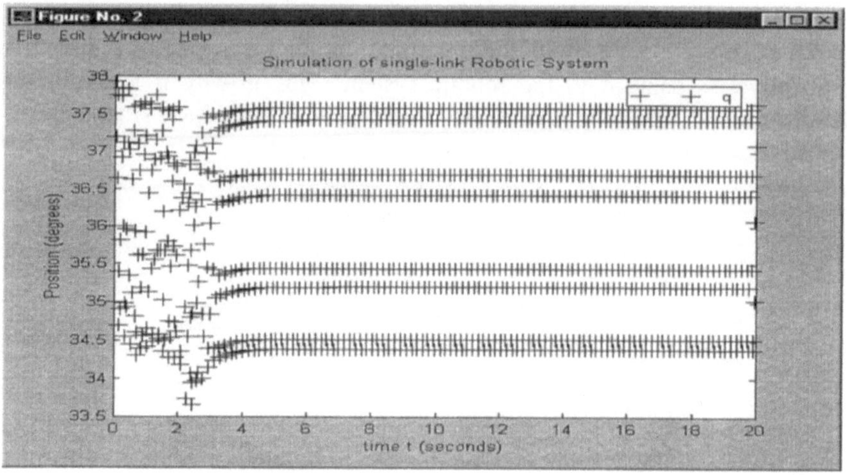

Figure 7.8 Simulation of a one-link robot manipulator.

7.4 Modelling Aircraft Dynamic Systems with the new Fuzzy Inference System

We describe in this section the application of our new fuzzy inference system to the problem of modelling aircraft dynamic systems. The mathematical models of aircraft systems can be represented as coupled non-linear differential equations (Melin & Castillo, 1998). In this case, we can develop a fuzzy rule base for model selection that enables the use of the appropriate mathematical model according to the changing conditions of the aircraft and its environment. For example, we can use the following model of an airplane when wind velocity is relatively small:

$$p' = I_1(-q + l) \qquad (7.16)$$
$$q' = I_2(p + m)$$

where I_1 and I_2 are the inertia moments of the airplane with respect to axis x and y, respectively, l and m are physical constants specific to the airplane, and p, q are the positions with respect to axis x and y, respectively. However, a more realistic model of an airplane in three dimensional space, is as follows:

$$p' = I_1(-qr + l) \qquad (7.17)$$
$$q' = I_2(pr + m)$$
$$r' = I_3(-pq + n)$$

where now I_3 is the inertia moment of the airplane with respect to the z axis, n is a physical constant specific to the airplane, and r is the position along the z axis. Considering now wind disturbances in the model, we have the following equation:

$$p' = I_1(-qr + l) - u_g \qquad (7.18)$$
$$q' = I_2(pr + m)$$
$$r' = I_3(-pq + n)$$

where u_g is the wind velocity. The magnitude of wind velocity is dependent on the altitude of the airplane in the following form:

$$u_g = u_{wind510} \left(1 + \frac{\ln(r/510)}{\ln(51)}\right) \qquad (7.19)$$

where $u_{wind510}$ is the wind speed at 510 ft altitude (typical value = 20 ft/sec).

If we use the general mathematical models of Equations (7.16)-(7.18) for describing aircraft dynamics, we can formulate a set of if-then rules that relate the models to the conditions of the aircraft and its environment. Lets assume that M_1 is given by Equation (7.16), M_2 is given by Equation (7.17), and M_3 is given by Equation (7.18). Now using the wind velocity u_g and inertia moment I_1 as selection parameters, we can establish the fuzzy rule base for model selection as in Table 7.2.

In Table 7.2, we are assuming that the wind velocity u_g can have only two possible fuzzy values (small and large). This is sufficient to know if we have to use the mathematical model that takes into account the effect of wind (M_3) for u_g large or if we don't need to use it and simply the model M_2 is sufficient (for u_g small). Also, the inertia moment (I_1) helps in deciding between models M_1 and M_2 (or M_3).

We show simulation results for an aircraft system obtained using our new method for modelling dynamical systems. In Figure 7.9 we show the simulation results for an airplane with inertia moments: $I_1 = 0.9$, $I_2 = 0.5$, $I_3 = 0.1$ and the physical constants are: $l = m = n = 0.1$. The initial conditions are the following: $p(0) = 0$, $q(0) = 0$, $r(0) = 0$.

We have implemented the new fuzzy inference system for multiple differential equations for modelling aircraft dynamic systems with very good results. We can conclude by saying that we consider our new method for modelling dynamical systems a good alternative for modelling complex real-world phenomena.

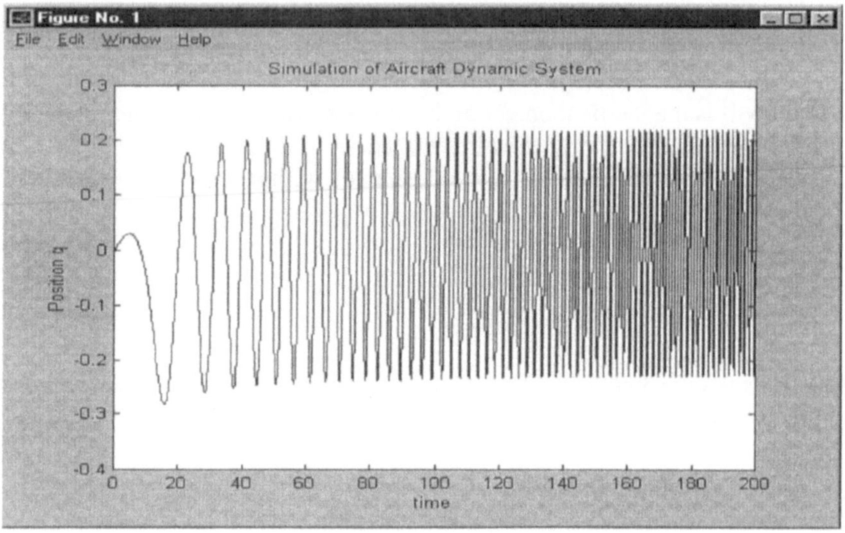

Figure 7.9 Simulation of position q for an airplane with $I_1 = 0.9$, $I_2 = 0.5$, $I_3 = 0.1$

Table 7.2 Fuzzy rule base for model selection of aircraft systems.

	IF	**THEN**
Wind Velocity, u_g	Inertia Moment, I_1	Mathematical Model
Small	Small	M_1
Small	Large	M_2
Large	Large	M_3

7. 5 Summary

We have developed a new fuzzy inference system for modelling complex dynamical systems using multiple differential equations. The new fuzzy inference system can be considered a generalization of the classical Sugeno fuzzy model. The use of differential equations as consequents instead of polynomials, gives to our new fuzzy modelling procedure more approximating power and as a result of this fact, a smaller number of fuzzy rules are needed to model a given problem.

We have illustrated in this paper our new method for modelling with the complex case of aircraft dynamic systems. Our new fuzzy inference system allows the change of mathematical models according to the changing conditions of the aircraft and its environment. Adequate modelling of complex dynamic systems (like the aircraft system) enables adaptive model-based control, which is a very important problem in real-world applications. We also illustrated our new method with the problem of modelling robotic dynamic systems. In this case, adequate modelling of these systems can help in achieve real time control, which is very important in real-world applications.

Chapter 8

A New Theory of Fuzzy Chaos for Simulation of Non-Linear Dynamical Systems

We describe in this chapter a new theory of chaos using fuzzy logic techniques. Chaotic behavior in non-linear dynamical systems is very difficult to detect and control. Part of the problem is that mathematical results for chaos identification are difficult to use in many cases, and even if one could use them there is an underlying uncertainty in the accuracy of the numerical simulations of the dynamical systems. For this reason, we can model the uncertainty of detecting the range of values where chaos occurs, using fuzzy set theory. Using fuzzy sets, we can build a new theory of Fuzzy Chaos, where we can use fuzzy sets to describe the behaviors of a system. We can also use fuzzy logic to build rules for behavior identification, using the mathematical knowledge from dynamical systems theory.

8.1 Problem Description

In the traditional mathematical theory of chaos, we have that 'chaotic behavior' is defined as sensitive dependence on initial conditions (Devaney, 1989). This sensitive dependence is measured with concepts like the Lyapunov exponents or the fractal dimension for the dynamical system. However, in the numerical simulations usually we have uncertainty related to numerical errors in the methods and also in the initial values (which are only approximated). For this reason, is very difficult to identify (in many cases) real chaotic behavior. The approach presented in this chapter is to relax the traditional mathematical definition of 'chaos' by using the theory of fuzzy logic (Zadeh, 1975), in this way obtaining a new more realistic definition of chaotic behavior. Our fuzzy chaos definition is a weaker definition of chaos, because we do not impose strict conditions on the accuracy of the numerical values for this behavior to occur. Also, it is an easier definition to apply to real world problems because in many cases we only have relative empirical evidence of chaos, which of course means that we have uncertainty about the identification of this behavior.

In this chapter, we propose several alternative definitions of fuzzy chaos and explore the advantages of each of these definitions. Also, we apply these definitions to the problems of modelling, simulation, and control of non-linear dynamical systems. In particular, we will consider the case of robotic dynamic systems, which is a very interesting problem from the point of view of the applications to manufacturing, but also it is important from the theoretical point of view. At the end, we also give fuzzy definitions for other types of dynamic behaviors because the identification of these behaviors can't be made without considering the associated uncertainty in predicting them. We will also show the implementation of these definitions as an intelligent system (developed in MATLAB) for simulation and behavior identification of robotic dynamic systems, to give an idea of the performance of this approach compared with the traditional one.

8.2 Towards a New Theory of Fuzzy Chaos

For a given dynamical system expressed as a non-linear differential equation:

$$dy/dt=f(t,y) \qquad y(0)=y_0 \tag{8.1}$$

Or as non-linear difference equation:

$$y_{t+1}=f(y_t,\ldots) \qquad y(0)=y_0 \tag{8.2}$$

We can have many different types of dynamic behaviors, for the above equations, depending on the parameter values and also, depending on the analytical properties of the function f. Also, there exists a fundamental difference between Equations (8.1) and (8.2), namely, that differential equations can only exhibit chaos when they are at least three-dimensional. However, difference equations can exhibit chaos even for the one-dimensional case.

In particular, we can have "chaotic behavior", defined formally as sensitive dependence on initial conditions for many real dynamical systems. However, in numerical simulations we usually have uncertainty related to numerical errors in the methods and also in the initial values. For this reason, is very difficult to identify precisely real chaotic behavior (Devaney, 1989). We can relax the traditional mathematical definition of "chaos" by using the theory of fuzzy logic (Zadeh, 1975), in this way obtaining a new more realistic definition of chaotic behavior. In a sense, we can say that our fuzzy chaos definition can be considered a weaker definition of chaos because we will require weaker conditions to identify this behavior.

We assume for the moment that we have a dynamical system in the real line given as:

$$y_t = f(y_{t-1}, \theta) \tag{8.3}$$

In this case, we can associate chaotic behavior with the number of period doublings (or bifurcations) that occur when the parameter θ is varied. According to this fact, we can state the following definition:

Definition 8.1 (chaotic behavior according to period doublings)
A one-dimensional dynamical system shows fuzzy chaos when the number of period doublings is considered to be large:

If number of period doublings is large **Then** behavior is fuzzy chaos.

Of course, other definitions can be made using other measures, like the fractal dimension or the Lyapunov exponents of the time series. For the case of the fractal dimension, we have extensive experimental results that show empirical evidence of chaos when the fractal dimension is close to a value of 2 in the plane. For this reason, we have proposed the following definition of fuzzy chaos:

Definition 8.2 (fuzzy chaos by the fractal dimension)
A one-dimensional dynamical system shows fuzzy chaos, when the value of the fractal dimension is large (close to a numeric value of 2 for the plane):

If the fractal dimension is large **Then** behavior is fuzzy chaos.

8.3 Fuzzy Chaos for Behavior Identification in the Simulation of Dynamical Systems

The simulation of Non-Linear Dynamical Systems consists in the iteration of a map or the numerical approximation of differential equations. In any case, the numerical simulation generates a time series:

$$y_1, y_2, \ldots, y_n. \tag{8.4}$$

After this, we have to perform behavior identification for the dynamical system. Here, we can apply the definition of fuzzy chaos to make a more efficient identification (because the theoretical definitions of chaos are difficult to apply). Of course, other types of behaviors can be defined similarly. For example, we can define the typical dynamic behaviors of a three-dimensional system with the following fuzzy rules:

If fractal dimension is low **Then** behavior is cycle of period 2
If fractal dimension is small **Then** behavior is cycle of period 4
If fractal dimension is regular **Then** behavior is cycle of period 8
If fractal dimension is medium **Then** behavior is cycle of period 16
If fractal dimension is high **Then** behavior is high order cycles
If fractal dimension is large **Then** behavior is fuzzy chaos.

Of course here the fractal dimension and the behavior are considered as linguistic variables and for a particular application the membership functions for the linguistic values will need to be defined. Also, the numerical value of the fractal dimension has to be calculated from the time series using the box dimension algorithm (Mandelbrot, 1987).

8.4 Simulation of Dynamical Systems

The problem of performing an efficient simulation for a particular dynamical system can be better understood if we consider a specific mathematical model. Let us consider the following model:

$$\begin{aligned} X' &= \sigma(Y-X) \\ Y' &= rX - Y - XZ \\ Z' &= XY - bZ \end{aligned} \tag{8.5}$$

where X, Y, Z, σ, r, b $\in$ R, and σ, r and b are three parameters which are normally taken, because of their physical origins, to be positive. The equations are often studied for different values of r in $0 < r < \infty$. This mathematical model, has been studied in (Rasband, 1990) to some extent, however there are still many questions to be answer for this model with respect to its very complicated dynamics for some ranges of parameter values.

If we consider simulating Equation (8.5), for example, the problem is of selecting the appropriate parameter values for σ, r, b, so that the interesting dynamical behavior of the model can be extracted. The problem is not an easy one, since we need to consider a three-dimensional search space σ r b and there are many possible dynamical behaviors for this model. In this case, the model consisting of three simultaneous differential equations, the behaviors can range from simple periodic orbits to very complicated chaotic attractors. Once the parameter values are selected then the problem becomes a numerical one, since then we need to iterate an appropriate map to approximate the solutions numerically.

The problem of performing automated simulation for a particular dynamical system is then of finding the "best" set of parameter values for the mathematical model. Our general algorithm (Castillo & Melin, 1995) for selecting the "best" set of parameter values is shown in Figure 8.1.

The algorithm shown in Figure 8.1 can be explained as follows. First, the mathematical model is analyzed to "understand" it. Second, a set of admissible parameters is generated using the understanding of the model. Third, a specific genetic algorithm is used to select the best set of parameter values. Finally, the numerical simulations are performed and the dynamical behaviors are identified using fuzzy logic .

STEP 1	Read the mathematical model M.
STEP 2	Analyze the model M to "understand" its complexity.
STEP 3	Generate a set of admissible parameters using the understanding of the model.
STEP 4	Perform a selection of the "best" set of parameter values. This set is generated using a specific genetic algorithm.
STEP 5	Perform the simulations by solving numerically the equations of the mathematical model. At this time the different types of dynamical behaviors are identified using a fuzzy rule base.

Figure 8.1 New algorithm for selecting the best set of parameter values.

The implementation of the new method for Automated Simulation as a computer program was done using the MATLAB programming language. The choice of MATLAB is because of its symbolic manipulation features and also because it is an excellent language for developing prototypes (Nakamura, 1997). The knowledge base for simulation consists of two modules: Parameter Selection, and Dynamic Behavior Identification. In the following lines we will describe these two modules in more detail.

8.5 Method for Automated Parameter Selection using Genetic Algorithms

The knowledge for simulation of the intelligent system consists in the application of a specific genetic algorithm (Goldberg, 1989) to select the best set of parameters of a particular dynamical system. Our genetic algorithm for parameter value selection (Castillo & Melin, 1998) can be defined as shown in Figure 8.2.

The fitness function should evaluate the dynamical information given by a particular set of parameter values, i.e. the fitness function should measure the power of the parameter set. Lets consider a three-dimensional model with 3 parameters θ, α, and γ, then assuming that we have only four possible dynamical behaviors (for a given system):

B0: fixed point of period 1
B1: fixed point of period 2
B2: fixed point of period 4
B3: fixed point of period 8
B4: chaotic behavior

STEP 1 Initialize a population with randomly generated individuals (parameters) and evaluate the fitness value of each individual
STEP 2 (a) Select two members from the population with probabilities proportional to their fitness values
(b) Apply crossover with a probability equal to the crossover rate
(c) Apply mutation with a probability equal to the mutation rate
(d) Repeat (a) to (d) until enough members are generated to form the next generation
STEP 3 Repeat steps 2 and 3 until the stopping criterion is met

Figure 8.2 Genetic algorithm for parameter value selection.

we will have that the parameter set $\pi = (\theta,\alpha,\gamma)$, where $\theta,\alpha,\gamma \in R$, can result in any of the five possible behaviors. In this case, we need to consider 5 individuals in the population and an initial population can be denoted as:

$$P_0 = (\pi_{01}, \pi_{02}, \pi_{03}\ \pi_{04}\ \pi_{05})$$

For an initial population there is a high probability that most of the π_i could give the B0 behavior, so there has to be evolution to obtain a better parameter set. The identification of the respective behaviors can be done by iteration of the dynamic systems or by other mathematical means, for example the fractal dimension or the Lyapunov exponents (Rasband, 1990). The fitness value of each individual in population P_i can by defined as follows:

$F(\pi_{ij}) = 1$ for fixed point of period 1
$F(\pi_{ij}) = 2$ for fixed point of period 2
$F(\pi_{ij}) = 4$ for fixed point of period 4
$F(\pi_{ij}) = 8$ for fixed point of period 8
$F(\pi_{ij}) = 10$ for chaotic behavior

this is only one of the possible schemes that can be used for this case. In this case, we have assigned the fitness values proportional to the complexity of the dynamic behavior to guide the search of the genetic algorithm. However, the specific numeric values could be changed to suit the needs of particular applications. We need to remember here that the fitness function has to be designed for each specific application of a genetic algorithm.

A more general form for defining the fitness function for real dynamical systems can be establish by using the fractal dimension (Mandelbrot, 1987) d_f of the time series generated by the numerical simulation of the dynamical system. Mathematically, we can define the fitness function as:

$$F(\pi_{ij}) = e^{df(\pi ij)}, \quad 0 \leq d_f \leq 3, \tag{8.6}$$

where $d_f(\pi_{ij})$ = fractal dimension of the time series for parameter set π_{ij}. The general idea of Equation (8.6) is to assign a bigger value to the fitness function when the complexity of the time series, generated by the simulation, is greater (which is true, of course, when d_f is of a higher value). Of course, here the use of the exponential function is only to spread the values of the fractal dimension but other functions could be used as well.

8.6 Method for Dynamic Behavior Identification using Fuzzy Logic

Once the parameter values have been found and the numerical simulations have been performed then the final step is to identify the possible dynamic behaviors of the system. The knowledge for behavior identification can be expressed as a fuzzy-rule base that uses the information obtained in the numerical simulation to identify the different behaviors of the model. To give an idea of how this knowledge can be expressed as a fuzzy-rule base we show below two sample schemes that can be used for behavior identification (Castillo & Melin, 1999).

8.6.1 Behavior Identification Based on the Analytical Properties of the Model

We can build a set of fuzzy rules for dynamic behavior identification based on the analytical properties of the mathematical models and using the well known theorems of dynamical systems theory (Castillo & Melin, 1997). To give an idea of how this knowledge can be translated to fuzzy rules we show below some sample rules for several types of dynamical systems.

1) Single-link Robot Model: This mathematical model, of a sinusoidally non-linear robot, consists of two simultaneous differential equations:

$$q' = Q \qquad (8.7)$$
$$Q' = (K_t I - N\sin(q) - F_d Q) / M_q$$

where the parameters I, M_q, N, F_d and K_t are all positive. Lim, Hu and Dawson (1996) have presented an extensive gallery of periodic and a-periodic motions for this model. In this case the equilibria (q*,Q*) is stable if and only if the real parts of the eigenvalues are negative and this is equivalent to the rule:

If $a > 0$ **Then** Equilibria = stable

where a is defined by the characteristic equation:

$$\lambda^2 + a\lambda + b = 0$$

with a = - trJ, b = detJ. Where "trJ" is the trace and "detJ" is the determinant of the Jacobian Matrix.

2) Other Bi-dimensional Models: Similar bi-dimensional autonomous models can be written in the following manner:

$$X' = \alpha\, f(X,Y) \qquad (8.8)$$
$$Y' = \beta\, g(X,Y)$$

In this case, the Equilibria (X*,Y*) is stable if:

$$\alpha\, f_x + (g_y - \beta) < 0$$

where f_x and g_y are partial derivatives. In fuzzy logic language we have the following rule:

If $[\alpha\, f_x + (g_y - \beta) < 0]$ **Then** Equilibria = stable

Also we have the following rule for a Hopf Bifurcation:

If $\alpha_0 = (\beta - g_y)/f_x$ **Then** Hopf_Bifurcation

which gives us the condition for a Hopf bifurcation to occur.

3) Firth's Model of a single-mode laser: The basic equations of a single-mode (unidirectional) homogeneously broadened laser in a high-finesse cavity, tuned to resonance, may be written as a system of three differential equations (Abraham & Firth, 1984):

$$\begin{aligned} X' &= \gamma_c (X + 2C_p) \\ P' &= -\Gamma (P - XD) \\ D' &= -\gamma (D + XP - 1) \end{aligned} \qquad (8.9)$$

Here X is a scaled electric field (or Rabi frequency), γ_c is a constant describing the decay of the cavity field and C is the cooperativity parameter.

In this case, the Equilibria (X*,P*,D*) is stable if a, b, c > 0 and (ab - c) > 0, where a, b and c are defined by the characteristic equation for the system. We can also have more complicated rules for other types of dynamical behaviors.

4) Other three-dimensional Models: A three-dimensional system of differential equations can be written in the following form:

$$\begin{aligned} X' &= \alpha f(X,Y,Z) \\ Y' &= \beta g(X,Y,Z) \\ Z' &= \gamma h(X,Y,Z) \end{aligned} \qquad (8.10)$$

In this case, the Equilibria (X*,Y*,Z*) is stable if a, b, c > 0 and (ab - c) > 0, where a, b and c are defined by the characteristic equation for the system:

$$\lambda^3 + a\lambda^2 + b\lambda + c = 0$$

In fuzzy logic language we have the rule:

If a,b,c >0 **And** (ab-c) >0 **Then** Equilibria = stable

other rules follow in the same manner for all the types of dynamical behaviors possible for this class of mathematical models.

We have to note here that in this case the computer program for this method needs to obtain the symbolic derivatives for the functions in the conditions of the rules. This is critical for the problem of behavior identification, since we require these derivatives to obtain the values of the parameters in the rules. This will make this method time consuming because the time series from the simulations are not used at all.

8.6.2 Behavior Identification Based on the Fractal Dimension and the Lyapunov Exponents

We can obtain a more efficient method of dynamic behavior identification, if we make use of the information contained in the time series that resulted from the simulation of the dynamical system. From the time series of the numerical simulations, we can calculate the Lyapunov exponents of the dynamical system and also the fractal dimension of the time series. With this dynamical information, we can easily identify the corresponding behaviors of the system.

For dissipative dynamical systems, we can use a set of fuzzy rules for behavior identification using the Lyapunov exponents. However, since the Lyapunov exponents can only identify between asymptotic stability, general limit cycles and chaos, we need to use the fractal dimension d_f to discriminate between the different periodic behaviors possible. Based on prior empirical work (Castillo & Melin, 1996), we have been able to use the fractal dimension to discriminate between different periodic behaviors. Then, if we combine the use of the Lyapunov exponents with the use of the fractal dimension, we can obtain a set of fuzzy rules that can identify in a one-to-one manner the different dynamic behaviors. The if-then rules have to be "fuzzy" because there is uncertainty associated with the numerical values of the Lyapunov exponents and also the classification scheme (for the limit cycles) using the fractal dimension is only approximated.

We show in Table 8.1 the knowledge base that we have developed for

Table 8.1 Knowledge base for behavior identification using Lyapunov exponents and fractal dimension

IF		THEN
Lyapunov exponents	Fractal Dim	Behavior Identification
(-)		stable fixed point
(-, -)		stable fixed point
(0, -)	[1.1, 1.2)	cycle period 2
(0, -)	[1.2, 1.3)	cycle period 4
(0, -)	[1.3, 1.4)	cycle period 8
(0, -)	[1.4, 1.5)	cycle period 16
(-, -, -)		stable fixed point
(0, -, -)	[2.1, 2.2)	cycle period 2
(0, -, -)	[2.2, 2.4)	cycle period 4
(0, -, -)	[2.4, 2.6)	cycle period 8
(0, -, -)	[2.6, 2.8)	cycle period 16
(+, 0, -)	[2.8, 3.0)	chaos

dynamic behavior identification for dynamical systems of up to three variables. The empty fields in Table 8.1 indicate no use of the fractal dimension for that case.

We can define membership functions for the numerical intervals of the fractal dimension, for the Lyapunov exponents and for the behavior identifications shown in Table 8.1. Once these membership functions are defined, the usual fuzzy reasoning methodology can be applied to implement this method of behavior identification. We show in the next section the implementation of this method in the fuzzy logic toolbox of MATLAB

8.7 Simulation Results for Robotic Systems

Our new method for automated simulation of non-linear dynamical systems was implemented as a prototype intelligent system in the MATLAB programming language. We tested the prototype intelligent system with different data to validate the new method and also the implementation with very good results. In this section, we show some of the results obtained using the intelligent system for automated simulation, to give an idea of the performance of the system. To give an idea of the performance of our fuzzy-fractal-genetic approach for simulation, we show below simulation results obtained for several types of dynamical systems. First, we show in Figure 8.3 the fuzzy rule base for the prototype intelligent system developed in the fuzzy logic toolbox of the MATLAB language.

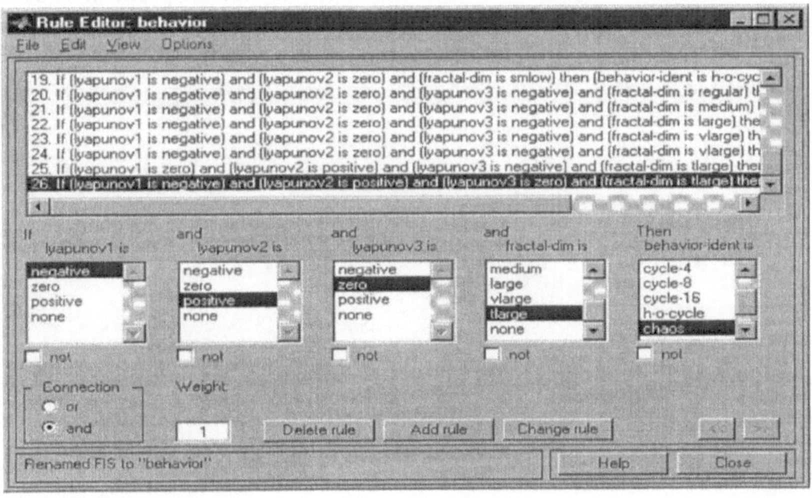

Figure 8.3 Fuzzy rule base in the rule editor.

We show in Figure 8.4 the membership functions for the linguistic values of the behavior identification. The membership functions were defined in the membership function editor of the fuzzy logic toolbox.

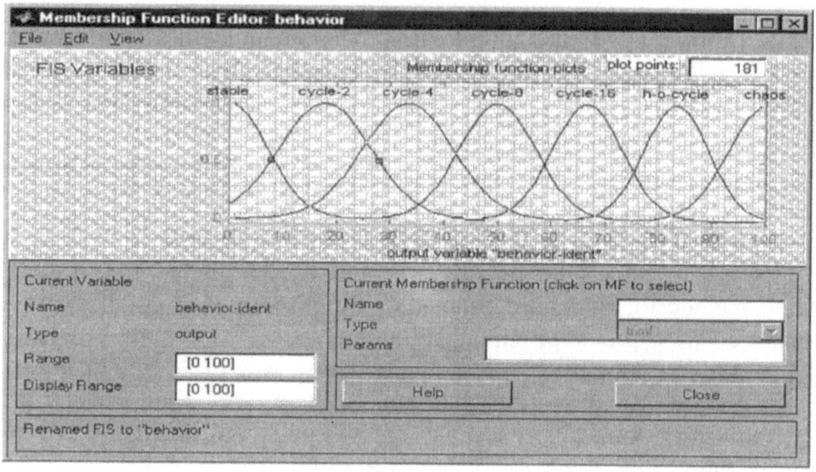

Figure 8.4 Membership function plots for the linguistic values of the behavior identification.

We show in Figure 8.5 the non-linear surface for the problem of behavior identification using as input variables Lyapunov exponents and fractal dimension.

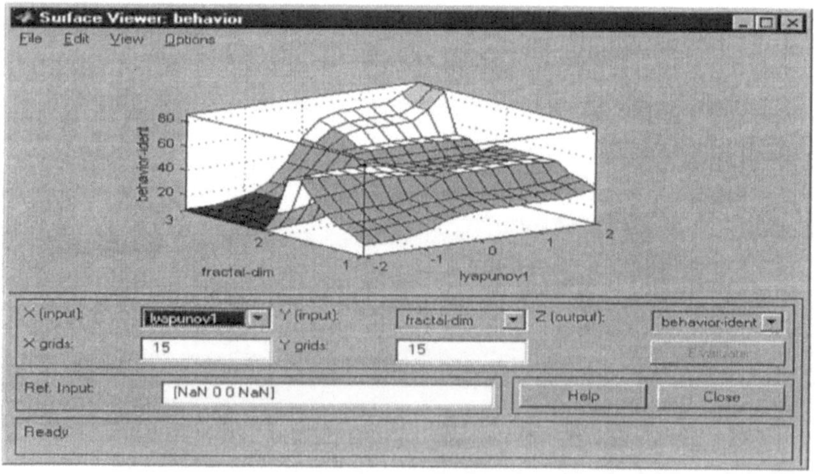

Figure 8.5 Non-linear surface for behavior Identification

We show in Figure 8.6 the reasoning procedure for behavior identification when specific values of the Lyapunov exponents and fractal dimension are given. In this figure, we can see how the final behavior identification for the dynamical system is evaluated with the Mamdani inference system.

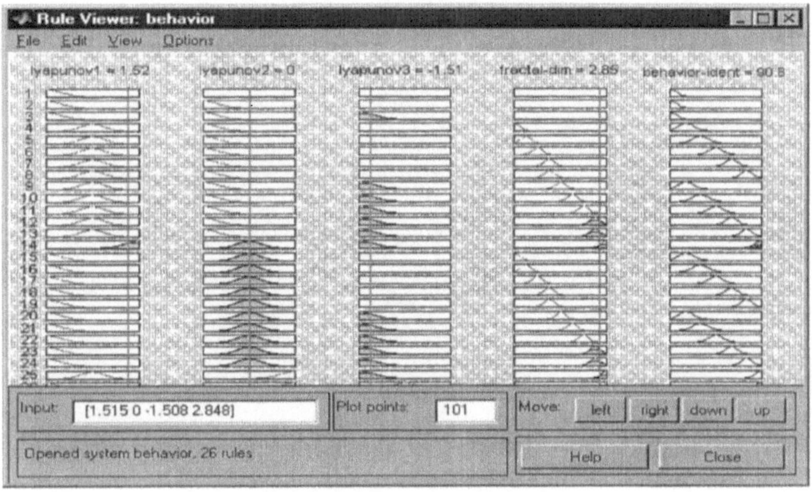

Figure 8.6 Behavior identification for specific values of Lyapunov exponents and fractal dimension

In Figure 8.7 we show the simulation results for a two-link robotic dynamic system with a mathematical model given by the two coupled second order differential equations:

$$q''_1 + a_1 q'_1 + b_1 \sin q_2 = c_1$$
$$q''_2 + a_2 q'_2 + b_2 \sin q_2 = c_2 \qquad (8.11)$$

where a_1, a_2, b_1, b_2, c_1 and c_2 are physical parameters of the robotic system. The simulation results shown in Figure 8.7 correspond to the parameter values: $a_1 = 212$, $a_2 = 20$, $b_1 = b_2 = 60$, $c_1 = c_2 = 72$ and to the initial conditions: $q_1(0) = 0.5$, $q'_1(0) = 5$, $q_2(0) = 5$, $q'_2(0) = 5$. The solution shown in Figure 8.7 is what is known as a "chaotic solution" because the orbit is oscillating (in an unstable manner) between an infinite number of periodic points. As a consequence of this fact the behavior identification in this case is of a "chaotic solution".

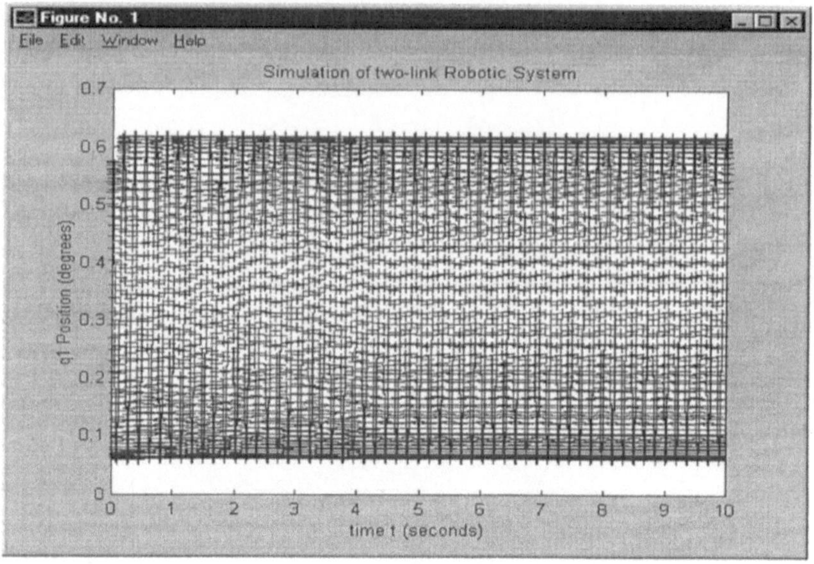

Figure 8.7 Simulation of a two-link robot arm showing chaotic behavior for position q_1.

8.8 Summary

We have presented in this chapter a new theory of Fuzzy Chaos for non-linear dynamical systems. We can apply this theory for behavior identification. We also presented in this chapter a new method for automated simulation of non-linear dynamical systems. This method is based on a hybrid fuzzy-fractal-genetic approach to achieve, in an efficient way, automated simulation for a particular dynamical system given its mathematical model. The use of genetic algorithms is to achieve automated parameter selection for the models. The use of fuzzy logic is to simulate the process of expert behavior identification by implementing the knowledge of identification by a set of fuzzy rules. We illustrated the application of this method for automated simulation for the case of robotic dynamic systems. The results show the efficiency of this new method for the simulation of complex non-linear dynamical systems. Of course, we need to apply our new approach for other types of dynamical systems, but we expect to get similar results.

Chapter 9

Intelligent Control of Robotic Dynamic Systems

We describe in this chapter a new method for adaptive model-based control of robotic dynamic systems using a new hybrid neuro-fuzzy-fractal approach. Intelligent control of robotic systems is a difficult problem because the dynamics of these systems is highly non-linear. We describe an intelligent system for controlling robot manipulators to illustrate our neuro-fuzzy-fractal hybrid approach for adaptive control. We use a new fuzzy inference system for reasoning with multiple differential equations for model selection based on the relevant parameters for the problem. In this case, the fractal dimension of a time series of measured values of the variables is used as a selection parameter. We use neural networks for identification and control of robotic dynamic systems.

9.1 Problem Description

Given the dynamic equations of motion of a manipulator, the purpose of robot arm control is to maintain the dynamic response of the manipulator in accordance with some pre-specified performance criterion (Fu, Gonzalez & Lee, 1987). Although the control problem can be stated in such a simple manner, its solution is complicated by inertial forces, coupling reaction forces, and gravity loading on the links. In general, the control problem consists of (1) obtaining dynamic models of the robotic system, and (2) using these models to determine control laws or strategies to achieve the desired system response and performance.

Among various adaptive control methods, the model-based adaptive control is the most widely used and it is also relatively easy to implement. The concept of model-based adaptive control is based on selecting an appropriate reference model and adaptation algorithm, which modifies the feedback gains to the actuators of the actual system.

Many authors have proposed linear mathematical models to be used as reference models in the general scheme described before. For example a linear second-order time invariant, differential equation can be used as the reference model for each degree of freedom of the robot arm. Defining the vector y(t) to represent the reference model response and the vector x(t) to represent the manipulator response, the joint i of the reference model can be described by

$$a_i y''_i(t) + b_i y'_i(t) + y_i(t) = r_i(t) \qquad (9.1)$$

If we assume that the manipulator is controlled by position and velocity feedback gains and the coupling terms are negligible, then the manipulator equation for joint i can be

$$\alpha_i(t)x''_i(t) + \beta_i(t)x'_i(t) + x_i(t) = r_i(t) \qquad (9.2)$$

where the system parameters $\alpha_i(t)$ and $\beta_i(t)$ are assumed to vary slowly with time.

The fact that this control approach is not dependent on a complex mathematical model is one of its major advantages, but stability considerations of the closed-loop adaptive system are critical. A stability analysis is difficult and has only been carried out using linearized models. However, the adaptability of the controller can become questionable if the interaction forces among the various joints are severe (non-linear). This is the main reason why soft computing techniques (Miller, Sutton & Werbos, 1995) have been proposed to control this type of dynamic systems.

9.2 Mathematical Modelling of Robotic Dynamic Systems

We will consider, in this section, the case of modelling robotic manipulators (Castillo & Melin, 1997). The general model for this kind of robotic system is the following:

$$M(q)q'' + V(q, q')q' + G(q) + F_d q' = \tau \qquad (9.3)$$

where $q \in R^n$ denotes the link position, $M(q) \in R^{nxn}$ is the inertia matrix, $V(q,q') \in R^{nxn}$ is the centripetal-Coriolis matrix, $G(q) \in R^n$ represents the gravity vector, $F_d \in R^{nxn}$ is a diagonal matrix representing the friction term, and τ is the input torque applied to the links.

For the simplest case of a one-link robot arm, we have the scalar equation:

$$M_q q'' + F_d q' + G(q) = \tau \tag{9.4}$$

If G(q) is a linear function (G = Nq), then we have the "linear oscillator" model:

$$q'' + aq' + bq = c$$

where $a = F_d/M_q$, $b = N/M_q$ and $c = \tau/M_q$. This is the simplest mathematical model for a one-link robot arm. More realistic models can be obtained for more complicated functions G(q). For example, if $G(q) = Nq^2$, then we obtain the "quadratic oscillator" model:

$$q'' + aq' + bq^2 = c \tag{9.5}$$

where a, b and c are defined as above.

A more interesting model is obtained if we define G(q) = Nsinq. In this case, the mathematical model is

$$q'' + aq' + b\sin q = c \tag{9.6}$$

where a, b and c are the same as above. This is the so-called "sinusoidally forced oscillator". More complicated models for a one-link robot arm can be defined similarly.

For the case of a two-link robot arm, we can have two simultaneous differential equations as follows:

$$\begin{aligned} q''_1 + a_1 q'_1 + b_1 q^2_2 &= c_1 , \\ q''_2 + a_2 q'_2 + b_2 q^2_1 &= c_2 \end{aligned} \tag{9.7}$$

which is called the "coupled quadratic oscillators" model. In Equation (9.7) a_1, b_1, a_2, b_2, c_1 and c_2 are defined similarly as in the previous models. We can also have the "coupled cubic oscillators" model:

$$\begin{aligned} q''_1 + a_1 q'_1 + b_1 q^3_2 &= c_1 , \\ q''_2 + a_2 q'_2 + b_2 q^3_1 &= c_2 \end{aligned} \tag{9.8}$$

9.3 Method for Adaptive Model-Based Control

Parametric Adaptive Control is the problem of controlling the output of a dynamical system with a known structure but unknown parameters. These parameters can be considered as the elements of a vector **p**. If p is known, the parameter vector θ of a controller can be chosen as θ^* so that the plant together with the fixed controller behaves like a reference model described by a differential equation with constant coefficients (Miller, Sutton & Werbos, 1995). If p is unknown, the vector $\theta(t)$ has to be adjusted on-line using all the available information concerning the dynamical system.

The structure of the adaptive system proposed in this work, to control a non-linear robotic dynamic system is similar to the one described in (Castillo & Melin, 1998), the main difference is that we use a decision scheme to select the appropriate reference model for the plant. The parameters of the neural network N_i are adjusted by backpropagating the identification error e_i while those of the neural network N_c are adjusted by backpropagating the control error (between the output of the reference model and the identification model) through the identification model.

9.3.1 Fuzzy Logic for Dynamic System Modelling

For a complex dynamical system (Rasband, 1990) it may be necessary to consider a set of models to represent adequately all of the possible dynamic behaviors of the system. We have designed a method (Castillo & Melin, 1999), based on fuzzy logic (Zadeh, 1975), for model selection using as input the numerical value of a selection parameter α. We assume, in what follows, that parameter α is defined over a real-valued interval:

$$\alpha_0 \leq \alpha \leq \alpha_n . \tag{9.9}$$

We also assume that we have n mathematical models considered appropriate for the respective n subintervals, defined on $[\alpha_0 , \alpha_n]$, as follows:

$$\alpha_0 \leq \alpha < \alpha_1 , \qquad \alpha_1 \leq \alpha < \alpha_2 , \ldots, \alpha_{n-1} \leq \alpha \leq \alpha_n . \tag{9.10}$$

The corresponding n mathematical models for these subintervals can be expressed as differential equations:

$$\begin{aligned} dy/dt &= f_1(y, \alpha) , \\ dy/dt &= f_2(y, \alpha) , \\ \ldots &\quad \ldots , \\ dy/dt &= f_n(y, \alpha) . \end{aligned} \tag{9.11}$$

Then, we can define a set of fuzzy if-then rules that basically relate the subintervals to the mathematical models in a one-to-one fashion. The advantage of using fuzzy rules (instead of conventional simple if-then rules) is that we can manage the underlying uncertainty of this process of model selection.

To implement this decision scheme, we need a reasoning method that can use differential equations as consequents. We have developed a new fuzzy inference system that can be considered as a generalization of Sugeno's inference system (Sugeno & Kang, 1988) in which we are now considering differential equations as consequents of the fuzzy rules, instead of simple polynomials. Using this method, the decision scheme can be expressed as a single-input fuzzy model as follows:

$$\begin{cases} \text{If } \alpha \text{ is small} & \text{then } dy/dt = f_1(y,\alpha) \\ \text{If } \alpha \text{ is regular} & \text{then } dy/dt = f_2(y,\alpha) \\ \quad\ldots & \quad\ldots \\ \text{If } \alpha \text{ is large} & \text{then } dy/dt = f_n(y,\alpha) \end{cases}$$

where the output y is obtained by the numerical solution of the corresponding differential equation. We have to note here that this new fuzzy inference system reduces to the standard Sugeno system only when the differential equations have closed-form solutions in the form of polynomials.

We describe below the reasoning procedure for our fuzzy inference system for the case of a one-input single-output fuzzy model. The procedure is very similar to the original Sugeno's procedure, except that now in the output we obtain the crisp values of "y" by solving numerically the corresponding differential equations. The numerical solutions of the differential equations can be achieved by the standard Runge-Kutta type method (Nakamura, 1997):

$$\begin{aligned} y_{n+1} &= RK(y_n) = y_n + 1/2(k_1 + k_2) \qquad (9.12) \\ k_1 &= hf(y_n, t_n) \\ k_2 &= hf(y_n + k_1, t_{n+1}) \end{aligned}$$

where h is the step size of the numerical method and RK can be considered as the Runge-Kutta operator that transforms numerical solutions from time n to time n+1.

The reasoning procedure for differential equations can also be used for rules with multiple inputs (for the case of several selection parameters) by simply considering the minimum of the firing strengths of each of the inputs.

9.3.2 Neuro-Fuzzy-Fractal Adaptive Model-Based Control

In this section, we combine the method for adaptive model-based control using neural networks (Melin & Castillo, 1998) with the method for model selection using fuzzy logic to obtain a new hybrid neuro-fuzzy method for control of non-linear dynamical systems. This new method combines the advantages of neural networks (ability for identification and control) with the advantages of fuzzy logic (ability for decision and use of expert knowledge) to achieve the goal of robust adaptive control of non-linear systems.

An intelligent control system designed with this method is capable of adapting to changing dynamic conditions in the plant, because it can change the control actions (given by the neural networks N_c) according to the data that is been measured on-line and also can change the reference mathematical model if there is a large enough change in the value of the selection parameter α. Of course, a change in the reference mathematical model also causes that the neural network N_i performs a new identification for the model. This is the reason why the whole process is called adaptive model-based control of non-linear dynamical systems.

This new method can be used for constructing intelligent control systems for different applications. This can be done by defining the appropriate set of mathematical models for the particular application (according to the type and complexity of the plant or system) and the correct architecture of the neural networks for identification and control. Initial training data can then be used to obtain the initial weights for the networks. The intelligent control system will then be ready for use on-line in the real plant or dynamical system. We have implemented a prototype intelligent control system, with the neuro-fuzzy approach for control, using the MATLAB© programming language.

9.4 Adaptive Control of Robotic Dynamic Systems

We use a fuzzy rule base for model selection for the case of robotic manipulators. We presented before mathematical models that can be used to model the dynamic behavior of robotic manipulators. Lets call M_1 the mathematical model given by Equation (9.5), M_2 the mathematical model given by Equation (9.6), M_3 the model given by Equation (9.7), and M_4 the model given by Equation (9.8). Then we can establish a fuzzy rule base for these models as explained before. We will assume here without loss of generality that the selection parameters are the fractal dimension of a time series of measured values of the relevant variables in the problem (angle, angular velocity) and the number of links of the manipulator. Also, we are assuming that only four models are needed to model completely the robotic system. Then, we can define a set of four fuzzy if-then rules that basically relate the fuzzy values of the selection parameters with the corresponding mathematical model. We show in Table 9.1 this set of fuzzy rules for model selection for the case of manipulators of one and two links.

We also need to define the membership functions for the fuzzy values in Table 9.1. The membership functions for the models should give us the degree of belief that a particular mathematical model is the correct one for the specific values of the selection parameters. We have to note here that for using a fuzzy rule base (like the one described in Table 9.1) with mathematical models, we need to use our new fuzzy inference system for multiple differential equations.

We use neural networks for identification and control of the robotic dynamic system. The neural networks are trained with the Levenberg-Marquardt algorithm with real data to achieve the desired level of performance. Two multilayer neural networks are used, one for identification of the model of the robotic system and the second for the controller. If we combine the fuzzy rule base for model selection with the neural networks for identification and control, we can obtain an intelligent system for adaptive model-based control of robotic dynamic systems.

Table 9.1. Fuzzy rule base for model selection of robotic systems

IF		THEN
Fractal dimension	Number of links	Mathematical Model
low	one	M_1
high	one	M_2
low	two	M_3
high	two	M_4

The intelligent control system combines the advantages of neural networks (ability for identification and control) with the advantages of fuzzy logic (use of expert knowledge) to achieve the goal of robust adaptive control of robotic dynamic systems. The general architecture of the intelligent control system for robotic systems is shown in Figure 9.1. In this figure, we have a module for the fuzzy-rule base of model selection, a module for the neural network of control, and a module for the neural network of identification.

An intelligent control system with the architecture shown in Figure 9.1 is capable of adapting to changing dynamic conditions in the robotic system, because it can change the control actions (given by the network Nc) according to the data measured on-line and also can change the reference mathematical model if there is a large enough change in the fractal dimension of the time series. Of course, a change in the reference mathematical model also causes that the neural network Ni performs a new identification for the model. In conclusion, the intelligent system with the architecture shown in Figure 9.1 achieves model-based control of robotic systems using a combination of Neural Networks and Fuzzy Logic.

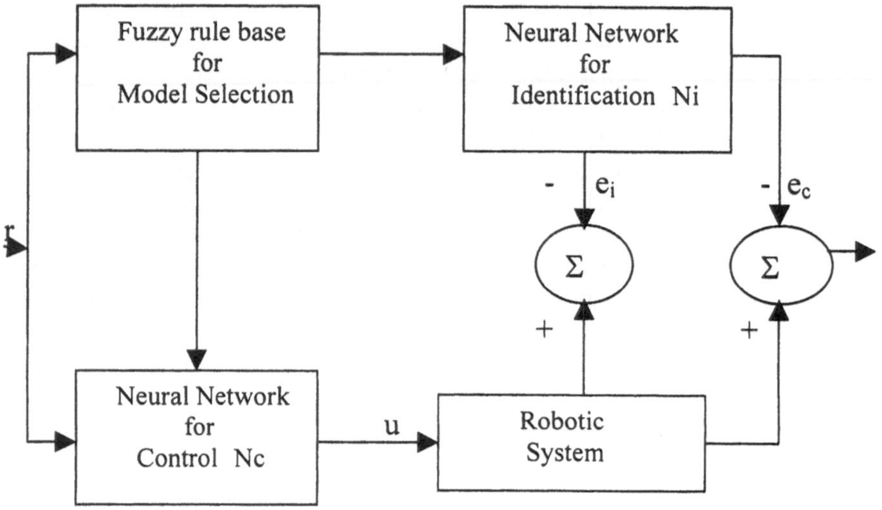

Figure 9.1. General architecture of the intelligent control system.

9.5 Simulation Results for Robotic Dynamic Systems

To give an idea of the performance of our neuro-fuzzy-fractal approach for adaptive model-based control of robotic systems, we show below simulation results obtained for a single-link robot arm. The desired trajectory for the link was selected to be

$$q_d = t\sin(2.0t) \tag{9.13}$$

and the simulation was carried out with the initial values: $q(0) = 0.1$ $q'_1(0) = 0$

We used three-layer neural networks (with 15 hidden neurons) with the Levenberg-Marquardt algorithm and hyperbolic tangent sigmoidal functions as the activation functions for the neurons. We show in Figure 9.2 the function approximation achieved with the neural network for control after 9 epochs of training with a variable learning rate. The identification achieved by the neural network can be considered very good because the error has been decreased to the order of 10^{-4}. We show in Figure 9.3 the curve relating the sum of squared errors SSE against the number of epochs of neural network training. We can see in this figure how the SSE diminishes rapidly from being of the order of 10^{2} to smaller value of the order of 10^{-4}. Still, we can obtain a better approximation by using more hidden neurons or more layers. In any case, we can see clearly how the neural networks learns to control the robotic system, because it is able to follow the arbitrary desired trajectory.

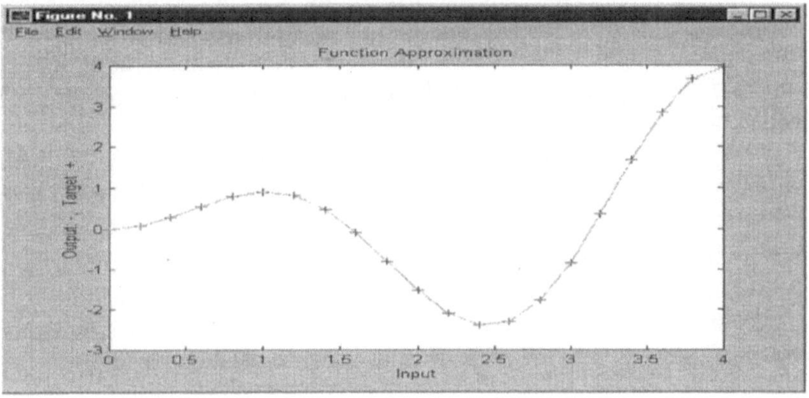

Figure 9.2 Function approximation after 9 epochs.

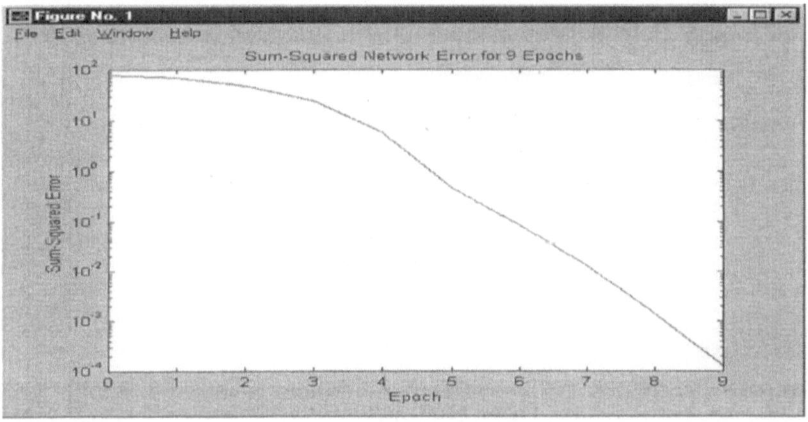

Figure 9.3 Sum of Squared Errors of the neural network.

We also show results for another desired trajectory for a one link robot arm. We used the following desired trajectory for the position of the arm

$$q_d = 1.0\sin(2.0(1\text{-}e^{-t^3})t)$$

and the simulation was carried out with the initial values:

$$q(0) = 0.1 \qquad q'_1(0) = 0$$

We used three-layer neural networks (with 10 hidden neurons) with the Levenberg-Marquardt algorithm and hyperbolic tangent sigmoidal functions as the activation functions for the neurons. We show in Figure 9.4 the initial function approximation achieved with the neural network for control. Of course, the approximation is not good (at the beginning) because the net hasn't been trained yet with the data.

We show in Figure 9.5 the function approximation achieved with the neural network for control after 400 epochs of training with a variable learning rate. The identification achieved by the neural network (after 400 epochs) can be considered very good because the error has been decreased to the order of 10^{-1}. Still, we can obtain a better approximation by using more hidden neurons or more layers. In any case, we can see clearly how the neural network learns to control the robotic system, so it is able to follow the arbitrary desired trajectory.

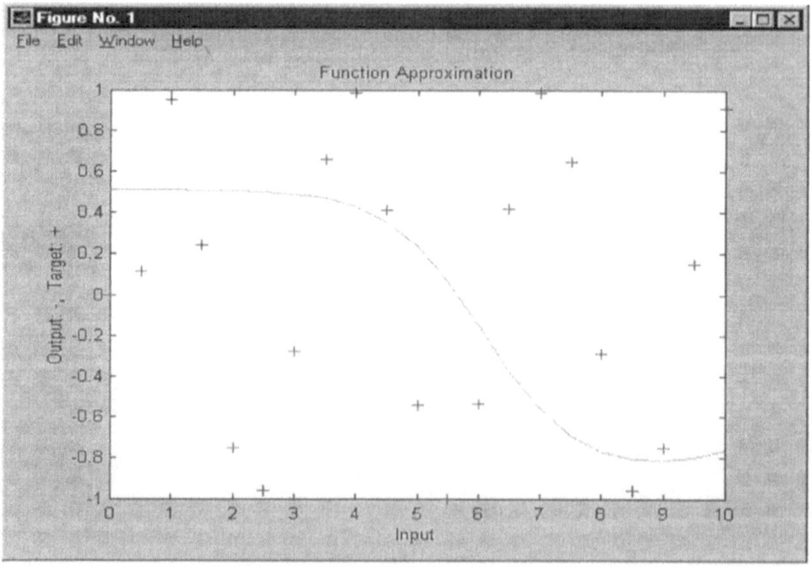

Figure 9.4. Initial function approximation of the neural network for control.

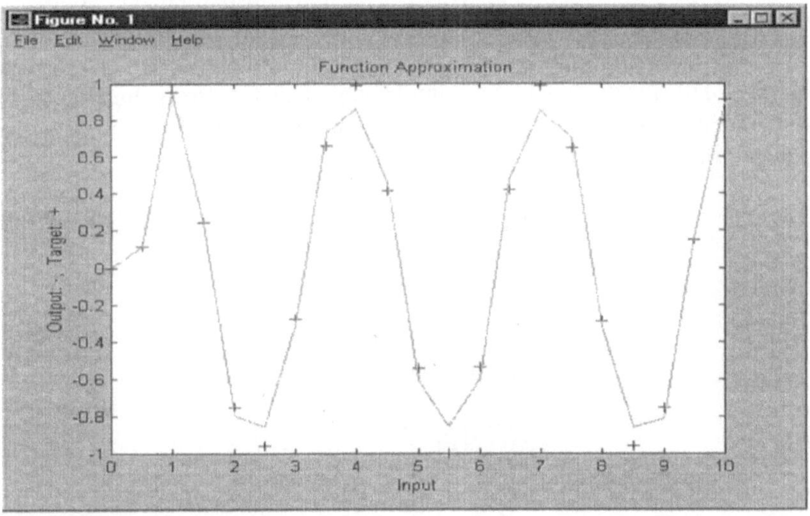

Figure 9.5. Function approximation of the neural network for control after 400 epochs.

We also show in Figure 9.6 the curve relating the sum of squared errors SSE against the number of epochs of neural network training. We can see in Figure 9.6 how the SSE diminishes rapidly from being of the order of 10^1 to a smaller value of the order of 10^{-1}.

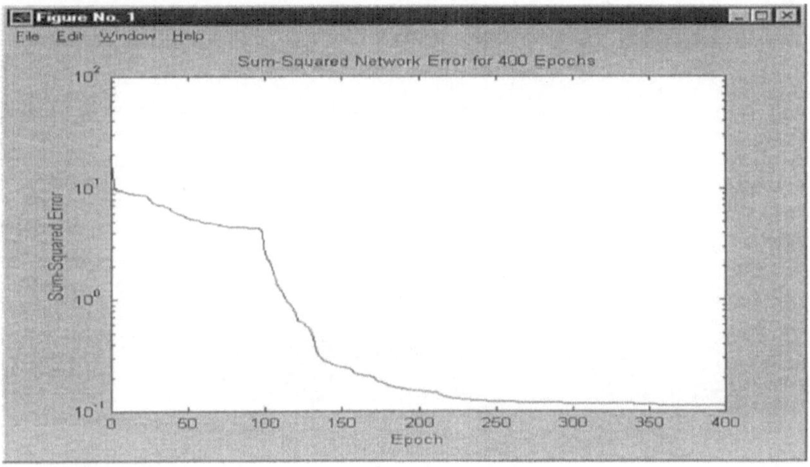

Figure 9.6 Sum of squares of errors for the neural network.

We have to mention here that these simulation experiments for a single link robot arm show very good results. We have also tried our approach for control with more complex robotic systems with encouraging results.

We show in Figure 9.7 the non-linear surface for the fuzzy rule base of Table 9.1. The fuzzy system was implemented in the fuzzy logic toolbox of MATLAB. We show in Figure 9.8 the reasoning procedure for specific values of the fractal dimension and number of links of the robotic system.

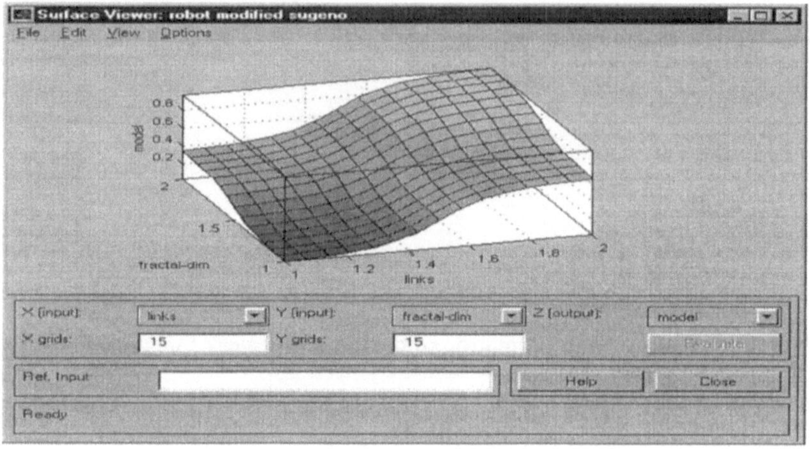

Figure 9.7 Non-linear surface for modelling,

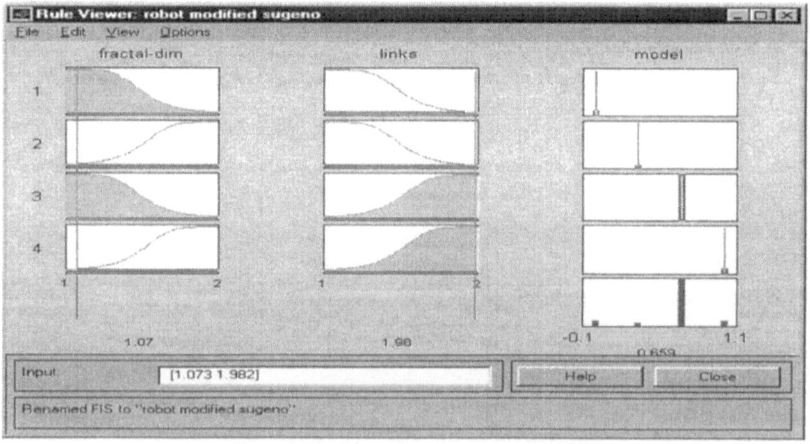

Figure 9.8 Fuzzy reasoning procedure

In Figures 9.9 and 9.10 we show simulation results for a two-link robot arm with a model given by two coupled second order differential equations. Figure 9.9 shows the behavior of position q1 and Figure 9.10 shows it for position q2 of the robot arm.

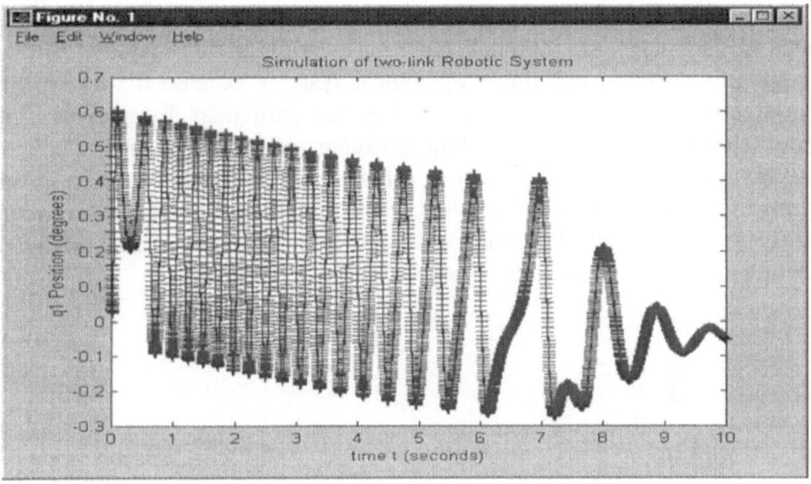

Figure 9.9 Simulation of position q1.

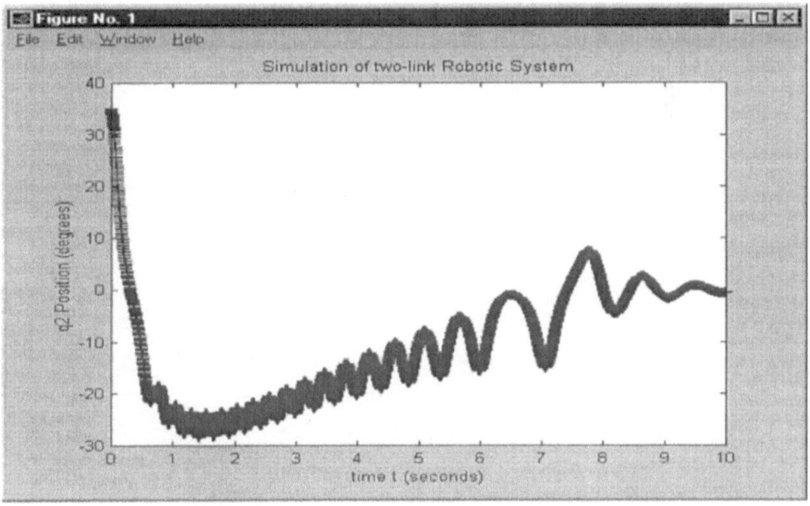

Figure 9.10 Simulation of position q2.

9.6 Summary

We have very good simulation results for several types of robotic systems for different conditions. The new method for control combines the advantages of neural networks (learning and adaptability) with the advantages of fuzzy logic (use of expert knowledge) to achieve the goal of robust adaptive control of robotic dynamic systems. We consider that our method for adaptive control can be applied to general non-linear dynamical systems because the hybrid neuro-fuzzy-fractal approach does not depend on the particular characteristics of the robotic dynamic systems. Controlling unstable and chaotic behavior in robotic dynamic systems is very important to achieve the desired performance in robots and also to avoid dangerous behavior of the system, which may cause physical damage to the system. We think that our contribution is a step in the right direction in achieving real time control for robotic systems.

Chapter 10

Controlling Biochemical Reactors

We describe in this chapter a general method for adaptive model-based control of non-linear dynamic plants using Neural Networks, Fuzzy Logic and Fractal Theory. The new neuro-fuzzy-fractal method combines Soft Computing (SC) techniques with the concept of the fractal dimension for the domain of Non-Linear Dynamic Plant Control. The new method for adaptive model-based control has been implemented as a computer program to show that our neuro-fuzzy-fractal approach is a good alternative for controlling non-linear dynamic plants. We illustrate in this chapter our new methodology with the case of controlling biochemical reactors in the food industry. For this case, we use mathematical models for the simulation of bacteria growth for several types of food. The goal of constructing these models is to capture the dynamics of bacteria population in food, so as to have a way of controlling this dynamics for industrial purposes. We use the fractal dimension for bacteria identification during the production process. We use neural networks for control and parameter identification, and fuzzy logic for modelling the complete dynamic system.

10.1 Introduction

We describe in this chapter a new method for adaptive control of non-linear dynamic plants based on the use of Neural Networks, Fuzzy Logic and Fractal Theory. Production processes in real world plants are often highly non-linear and difficult to control (Melin & Castillo, 1996). The problem of controlling them using conventional controllers has been widely studied (Albertos, Strietzel & Mart, 1997). Much of the complexity in controlling any process comes from the complexity of the process being controlled. This complexity can be described in several ways. Highly non-linear systems are difficult to control, particularly when they have complex dynamics (such as instabilities to limit cycles and chaos). Difficulties can often be presented by constraints, either on the control parameters or in the operating regime. Lack of exact knowledge of the process, of course, makes control more difficult. Optimal control of many processes also requires systems, which make use of predictions of future behavior. The mathematical models for the plants are assumed to be systems of differential equations. The goal of having these models is to capture the dynamics of production processes, so as to have a way of controlling this dynamics for industrial purpose (Melin & Castillo, 1997).

We need a mathematical model of the non-linear dynamic plant to understand the dynamics of the processes involved in production. For a specific case, this may require testing several models before obtaining the appropriate mathematical model for the process (Melin & Castillo, 1998). For real world plants with complex dynamics, we may even need several models for different sets of parameter values to represent all of the possible behaviors of the plant. The general mathematical model of a plant can be expressed as follows:

$$dx/dt = f_1(x, D, \alpha) - \beta f_2(x, D, \alpha) \qquad (10.1)$$
$$dp/dt = \beta f_2(x, D, \alpha)$$

where $x \in R^n$ is a vector of state variables, $p \in R^m$ is a vector of products, $\beta \in R$ is a constant measuring the efficiency of the conversion process, $D \in (0, 3)$ is the fractal dimension of the process, and $\alpha \in R$ is a selection parameter. The fractal dimension is used to characterize the production process, for example in the case of biochemical reactors D represents the fractal dimension of the bacteria used for production (Melin & Castillo, 1998).

For a complex dynamical system it may be necessary to consider a set of mathematical models to represent adequately all of possible dynamic behavior of the system. In this case, we need a decision scheme to select the appropriate model to use according to the linguistic value of a selection parameter α. We use a new fuzzy inference system for differential equations to achieve model selection. We have fuzzy rules of the form:

IF α is A_1 **AND** D is B_1 **THEN** M_1 (10.2)

... ...

IF α is A_n **AND** D is B_n **THEN** M_n

where $A_1, \ldots, A_n$ are linguistic values for α, $B_1, \ldots, B_n$ are linguistic values for the fractal dimension D, and $M_1, \ldots, M_n$ are mathematical models of the form given by equation (1). The selection parameter α can be the temperature for biochemical processes, because temperature changes cause the presence of new bacteria in this case (Melin & Castillo, 1998).

We combine adaptive model-based control using neural networks with the method for model selection using fuzzy logic and fractal theory, to obtain a new hybrid neuro-fuzzy-fractal method for control of non-linear plants. This general method combines the advantages of neural networks (ability for identification and control) with the advantages of fuzzy logic (ability for decision and use of expert knowledge) to achieve the goal of robust adaptive control of non-linear dynamic plants. We also use the fractal dimension to characterize the production processes in modelling these dynamical systems. We have developed intelligent control systems using this new method for adaptive control for several applications, to validate our new approach for control. We have obtained very good results in controlling biochemical reactors and chemical reactors with the hybrid approach for control.

10.2 Fuzzy Logic for Modelling

We have designed a method, based on fuzzy logic techniques, for mathematical model selection using as input the numerical value of a selection parameter α. We assume, in what follows, that parameter α is defined over a real-valued interval:

$$\alpha_0 \leq \alpha \leq \alpha_n . \qquad (10.3)$$

We also assume that we have n mathematical models considered the most appropriate ones for the respective n subintervals, defined on $[\alpha_0 , \alpha_n]$, as follows:

$$\alpha_0 \leq \alpha < \alpha_1 , \quad \alpha_1 \leq \alpha < \alpha_2 ,\ldots, \quad \alpha_{n-1} \leq \alpha \leq \alpha_n . \qquad (10.4)$$

The corresponding n mathematical models for these subintervals can be expressed as differential equations:

$$\begin{aligned} dy/dt &= f_1(y, \alpha) \qquad (10.5) \\ dy/dt &= f_2(y, \alpha) \\ &\ldots \\ dy/dt &= f_n(y, \alpha) \end{aligned}$$

Then, we can define a set of fuzzy if-then rules that basically relate the subintervals to the mathematical models in a one-to-one fashion. The advantage of using fuzzy rules (instead of conventional simple if-then rules) is that we can manage the underlying uncertainty of this process of model selection. We show the basic decision scheme for developing a fuzzy rule base for model selection in Table 10.1.

Table 10.1 Decision scheme for model selection.

IF	THEN
$\alpha_0 \le \alpha < \alpha_1$	M_1: $dy/dt = f_1(y, \alpha)$
$\alpha_1 \le \alpha < \alpha_2$	M_2: $dy/dt = f_2(y, \alpha)$
$\alpha_2 \le \alpha < \alpha_3$	M_3: $dy/dt = f_3(y, \alpha)$
...	...
$\alpha_{n-1} \le \alpha \le \alpha_n$	M_n: $dy/dt = f_n(y, \alpha)$

Of course, for this decision scheme to work we need to define membership functions for the corresponding mathematical models. The membership functions for the models should give us the degree of belief that a particular model is the correct one for a specific value of the parameter α.

To apply this method of model selection, to a particular application, we have to find the corresponding selection parameter α to be used in the decision scheme proposed in Table 10.1. Then, a partition of the definition interval for α has to be performed. After this, the one-to-one map between the mathematical models and the subintervals (obtained from the partition) is constructed. In this way, we can obtain the fuzzy rule base for model selection for a particular application.

10.3 Neural Networks for Control

Parametric Adaptive Control is the problem of controlling the output of a system with a known structure but unknown parameters. These parameters can be considered as the elements of a vector p. If p is known, the parameter vector θ of a controller can be chosen as θ^* so that the plant together with the fixed controller behaves like a reference model described by a difference (or differential) equation with constant coefficients (Narendra & Annaswamy, 1989). If p is unknown, the vector $\theta(t)$ has to be adjusted on-line using all the available information concerning the system.

Two distinct approaches to the adaptive control of an unknown plant are (i) direct control and (ii) indirect control. In direct control, the parameters of the controller are directly adjusted to reduce some norm of the output error. In indirect control, the parameters of the plant are estimated as p(t) at any time instant and the parameter vector $\theta(t)$ of the controller is chosen assuming that p(t) represents the true value of the plant parameter vector. Even when the plant is assumed to be linear and time-invariant, both direct and indirect adaptive control results in non-linear systems.

When indirect control is used to control a non-linear system, the plant is parameterized using a mathematical model of the general form described in section 1 and the parameters of the model are updated using the identification error. The controller parameters in turn are adjusted by backpropagating the error (between the identified model and the reference model outputs) through the identified model. A block diagram of such an adaptive system is shown in Figure 10.1.

The overall structure of the adaptive system proposed in this paper to control a non-linear dynamical system is the same as shown in Figure 10.1 and is independent of the specific model used to identify the plant. The delayed values of the plant input and plant output form the inputs to the neural network Nc which generates the feedback control signal to the plant. The parameters of the Neural Network Ni are adjusted by backpropagating the identification error e_i while those of the Neural Network Nc are adjusted by backpropagating the Control error (between the output of the reference model and the identification model) through the identification model.

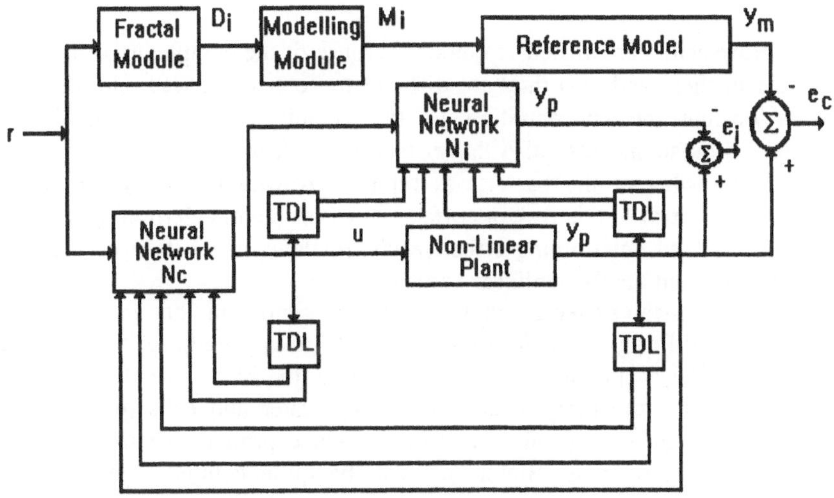

Figure 10.1 Indirect Adaptive Neuro-Fuzzy-Fractal Control.

The mathematical model for the non-linear plant in the time domain is generated by the method of modelling (described in section 10.2) using the real data that is measured on-line in the plant. On the other hand, the fractal module is used to characterize the production process and this information is used to specify the mathematical model in the time and space domain. This scheme enables the dynamic changes of models according to the changes of on-line process identification. Our new method for adaptive model-based control combining Neural Networks, Fuzzy Logic and Fractal Theory differs from our previous approach of considering only the use of neural networks and models (Melin & Castillo, 1997).

10.4 Adaptive Control of a Non-Linear Plant

Process control of biochemical plants is an attractive application because of the potential benefits to both adaptive network research and to actual biochemical process control. In spite of the extensive work on self-tuning controllers and model-reference control, there are many problems in chemical processing industries for which current techniques are inadequate. Many of the limitations of current adaptive controllers arise in trying to control poorly modeled non-linear systems (Albertos, Strietzel & Mart, 1997). For most of these processes extensive data are available from past runs, but it is difficult to formulate precise models. This is precisely where adaptive networks are expected to be useful (Ungar, 1995).

Bioreactors are difficult to model because of the complexity of the living organisms in them and also they are difficult to control because one often can't measure on-line the concentration of the chemicals being metabolized or produced. Bioreactors can also have markedly different operating regimes, depending on whether the bacteria is rapidly growing or producing product. Model-based control of this reactors offers a dual problem: determining a realistic process model and determining effective control laws in the face of inaccurate process models and highly nonlinear processes.

Biochemical systems can be relatively simple in that they have few variables, but still very difficult to control due to strong nonlinearities which are difficult to model accurately. A prime example is the bioreactor. In its simplest form, a bioreactor is simply a tank containing water and cells (e.g.. bacteria) which consume nutrients ("substrate") and produce products (both desired and undesired) and more cells. Bioreactors can be quite complex: cells are self-regulatory mechanisms, and can adjust their growth rates and production of different products radically depending on temperature and concentrations of waste products. Systems with heating or cooling, multiple reactors or unsteady operation greatly complicate the analysis. Mathematical models for these systems can be expressed as differential (or difference) equations.

Now we propose mathematical models that integrate our method for geometrical modelling of bacteria growth using the fractal dimension (Castillo & Melin, 1994) with the method for modelling the dynamics of bacteria population using differential equations (Melin & Castillo, 1998). The resulting mathematical models describe bacteria growth in space and in time, because the use of the fractal dimension enables us to classify bacteria by the geometry of the colonies and the differential equations help us to understand the evolution in time of bacteria population.

We will consider first the case of using one bacteria for food production. The mathematical model in this case can be of the following form:

$$dN/dt = r(1 - N^{-D}/K)N^{-D} - \beta N^{-D}$$
$$dP/dt = \beta N^{-D} \qquad (10.6)$$

where D is the fractal dimension, N is the bacteria population, P is quantity of chemical product, r is the rate of bacteria growth, K is the environment capacity, and β is a biochemical conversion factor.

We will consider now the case of two bacteria used for food production:

$$dN_1/dt=[r_1-(r_1/K_1)N_1^{-D1}-(r_1/K_1)\delta_{12}N_2^{-D2}]N_1^{-D1}-\beta N_1^{-D1}$$
$$dN_2/dt = [r_2-(r_2/K_2)N_2^{-D2}-(r_2/K_2)\delta_{21}N_1^{-D1}]N_2^{-D2}-\gamma N_2^{-D2}$$
$$dP/dt = \beta N_1^{-D1} + \gamma N_2^{-D2} \qquad (10.7)$$

where D_1 is the fractal dimension of bacteria 1, D_2 is the fractal dimension of bacteria 2 and the rest of variables are as described in the last equation.

As we can see from equations (10.6) and (10.7) the idea of our method of modelling is to use the fractal dimension D as a parameter in the differential equations, so as to have a way of classifying for which type of bacteria the equation corresponds. In this way, equation (10.6), for example, can represent the model for food production using one bacteria (the one defined by the fractal dimension D).

We have implemented a model-based neural controller using the architecture of Figure 10.1. Two multilayer networks are used, one for the model of the plant and the second for the controller. Each network at the moment has 2 layers of 5 hidden nodes; more nodes will, of course, give better accuracy. The Neural Networks were implemented in the MATLAB programming language to achieve a high level of efficiency on the numerical calculations needed for these modules. The Fractal module was also implemented in the MATLAB programming language for the same reason. In this way we combine the three methodologies to obtain the best of the three worlds (Neural Networks, Fuzzy Logic and Fractal Theory) using for each the appropriate implementation language.

10.5 Fractal Identification of Bacteria

In this section we describe a method for the identification of microorganisms that enables quality control for the food production process. This Quality Control is done by monitoring the types of Bacteria present in samples of food extracted from production lines. We use a method for the identification of microorganisms developed by the authors (Castillo & Melin, 1994), based on the use of the fractal dimension, to eliminate the need of applying a long sequence of microbiological techniques to the samples of food.

This method uses the fractal dimension to make a unique classification of the different types of microorganisms, because it is a known experimental fact that the colonies of different types of bacteria have different geometrical forms. The problem is then to finding a one to one map between the different types of bacteria and their corresponding fractal dimension, in this way obtaining a unique method of identification of microorganisms for the food industry. The first step in obtaining this map is to find experimentally (in the lab) the different geometrical forms for the bacteria. The second step is to calculate the corresponding fractal dimensions for the different types of bacteria. This fractal dimension can be calculated for a selected type of bacteria with several samples, to obtain as a result a statistical estimation of the dimension and the corresponding error of the estimation. In order to make an efficient use of this map between the different types of bacteria and their corresponding estimated dimensions, we need to implement it as a module in the computer program.

The method for the identification of microorganisms using the fractal dimension can be stated mathematically in the following form: let M_m be a one to one map between the sets I_m and D_m, where the set I_m can be called set of identifications of microorganisms and the set D_m can called set of fractal dimensions. The set of identifications can be as follows:

$$I_m = \{\text{staphylococcus}_{\text{aureus}}, \text{streptococcus}_{\text{fecalis}}, \text{pseudomona}_{\text{aureuginosa}}, \text{salmonella}_{\text{typhi}}, \ldots\}$$

and the set of fractal dimensions can be as follows:

$$D_m = \{D_{sa}, D_{sf}, D_{pa}, D_{st}, \ldots\}$$

where D_{sa} is the fractal dimension of the Staphylococcus aureus bacteria, D_{sf} is the fractal dimension of the Streptococcus fecalis bacteria, D_{pa} is the fractal dimension of the Peseudomona aureuginosa, D_{st} is the fractal dimension of the Salmonella typhi and so on. In all of the above the fractal dimension is given by:

$$D = [\log(N)/\log(1/r)] \tag{10.8}$$

for an object of N parts each scaled down by a ratio, r. For an estimation of the fractal dimension we can use the equation:

$$N(r) = \beta \, 1/r^{D} \qquad (10.9)$$

where N(r) = number of boxes contained in a piece of object and r = size of the box. Counting the number of boxes for different sizes and performing a linear logarithmic regression we can estimate the box dimension of an object with the equation:

$$\ln N(r) = \ln\beta - D\ln r \qquad (10.10)$$

The authors have implemented this method as a computer program to obtain an automated method of identification of bacteria (Castillo & Melin, 1994). In this case this method is used to verify the types of bacteria that we have at any moment in our industrial microbiology process, in this way controlling the types of bacteria in the process. This is important for quality control because in some cases harmful bacteria can appear during the production.

10.6 Experimental Results

We describe in this section some of the experimental results that we have achieved with the intelligent system for adaptive neuro-fuzzy-fractal control. In Table 10.2 we show the training data and results obtained with the neural network for identification (Ni). The neural network identification has the goal of simulating the mathematical model of the non-linear dynamic plant.

Table 10.2.- Training data and results for Ni.

Training method	Learning rate	Number epochs	Math Model	Initial Cond.	Final error
BP	0.001	80,000	M2	P=0 N_1=26.5 N_2=26.5	10^{3}
BP	0.000001	120,000	M2	P=0 N_1=26.5 N_2=26.5	10^{2}
BP	0.0001	40,000	M2	P=0 N_1=26.5 N_2=26.5	10^{1}
BP	0.0001	80,000	M2	P=0 N_1=26.5 N_2=26.5	10^{-2}

From Table 10.2, we can see that we achieved very good identification results with network used for a learning rate of 0.0001 (In all cases, we denote the model of one bacteria as M1, the model of two bacteria as M2 and the one of three bacteria as M3). In Table 10.3 we show some of the numerical simulations of the dynamical behavior of the typical biochemical reactors.

We can see from Table 10.3 that there is a wide range of possible dynamical behaviors for biochemical reactors. This makes the problem of controlling these reactors very difficult. The results obtained for the Neural Network of Control (Nc) were similar than the ones achieved for identification. The errors of control achieved are relatively small and can be considered very good for this application.

Table 10.3.-Simulations of the dynamical behavior of biochemical reactors

Temp.	Math Model	Initial Cond.	Final Product	Limiting dynamical behavior
102 °F	M1	N=97.5	70 %	cycle of period 8
110 °F	M2	N_1=26.5 N_2=26.5	80 %	cycle of period 16
115 °F	M3	N_1=65 N_2=6.5 N_3=10	60 %	fixed points: N_1=20 N_2 =0 N_3=120
118 °F	M3	N_1=60 N_2=60 N_3=0.5	65 %	fixed points: N_1=0 N_2 =190 N_3=70

We show in Figure 10.2 simulation results of bacteria population used for food production. We can see from this figure the complicated dynamics for the case of two bacteria competing in the same environment, and at the same time producing the chemical product necessary for food production.

We also show in Figure 10.3 simulation results for the case of two good bacteria used for food production and one bad bacteria that is attacking the other ones. We can see from this figure how one of the good bacteria is eliminated (the population goes down to zero), which of course results in a decrease of the resulting quantity of the food product. This is a case, which has to be avoided because of the bad resulting effect of the bad bacteria. Intelligent control helps in avoiding these types of scenarios for food production.

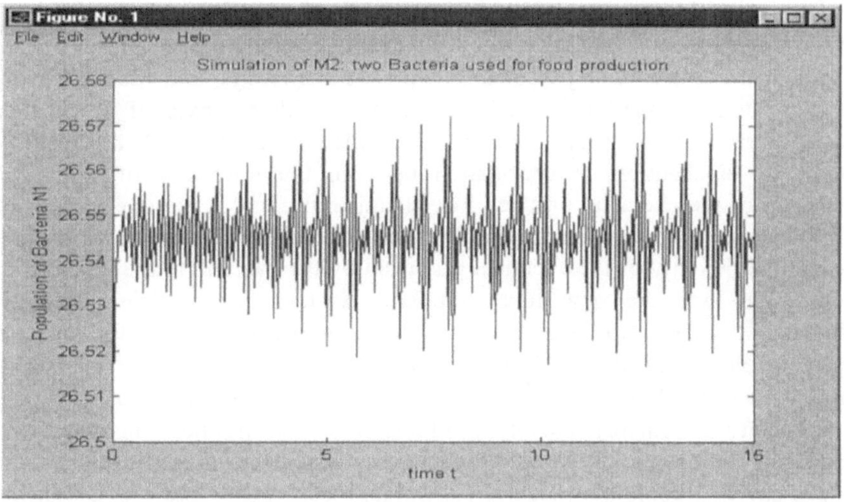

Figure 10.2 Simulation of the model for two bacteria used in food production.

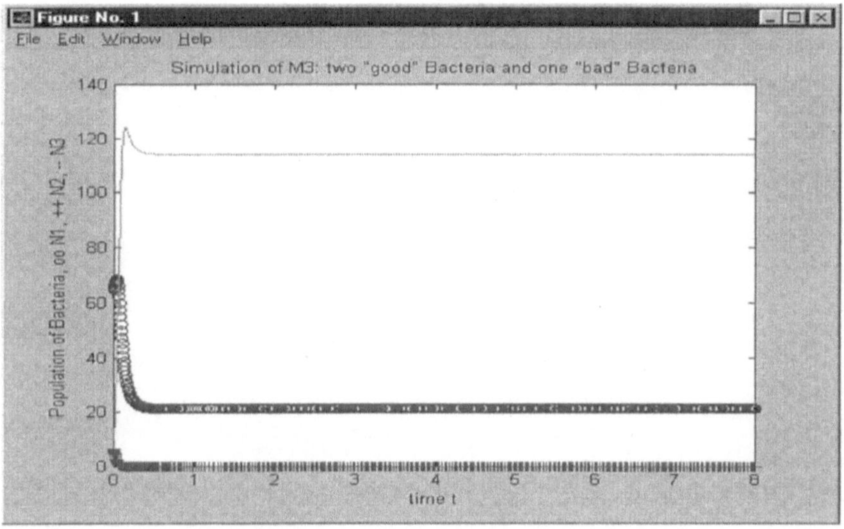

Figure 10.3 Simulation of the model for two good bacteria and one bad bacteria.

10.7 Summary

We have developed a general method for adaptive model based control of non-linear dynamic plants using Neural Networks, Fuzzy Logic and Fractal Theory. We illustrated our method for adaptive control with the case of biochemical reactors in the food industry. In this case, the mathematical models represent the process of biochemical transformation between the microbial life and their generation of the chemical product. The chemical product is the one responsible for producing the food in the process. So, the goal here is to increase as much as possible the quantity of the chemical produced by the bacteria in the biochemical reactor. We used the new fuzzy inference system for multiple differential equations to take into consideration the appropriate models for different conditions in the biochemical reactor. We also described in this chapter an adaptive controller based on the use of neural networks and mathematical models for the plant (the reactor). The proposed adaptive controller performs rather well considering the complexity of the domain being considered in this research work. We can say that combining Neural Networks, Fuzzy Logic and Fractal Theory, using the advantages that each of these methodologies has, can give good results for this kind of application. Also, we believe that our neuro-fuzzy-fractal approach is a good alternative for solving similar problems.

Chapter 11

Controlling Aircraft Dynamic Systems

We describe in this chapter a hybrid method for adaptive model-based control of non-linear dynamic systems using Neural Networks, Fuzzy Logic and Fractal Theory. The new neuro-fuzzy-fractal method combines Soft Computing (SC) techniques with the concept of the fractal dimension for the domain of Non-Linear Dynamic System Control. The new method for adaptive model-based control has been implemented as a computer program to show that our neuro-fuzzy-fractal approach is a good alternative for controlling non-linear dynamic systems. It is well known that chaotic and unstable behavior may occur for non-linear systems. Normally, we will need to control this type of behavior to avoid structural problems with the system. We illustrate in this chapter our new methodology with the case of controlling aircraft dynamic systems. For this case, we use mathematical models for the simulation of aircraft dynamics during flight. The goal of constructing these models is to capture the dynamics of the aircraft, so as to have a way of controlling this dynamics to avoid dangerous behavior of the aircraft dynamic system.

11.1 Introduction

We describe in this chapter a new method for adaptive control of non-linear dynamic systems based on the use of Neural Networks, Fuzzy Logic and Fractal Theory. The dynamics of real world systems are often highly non-linear and difficult to control (Melin & Castillo, 1996). The problem of controlling them using conventional controllers has been widely studied (Albertos, Strietzel & Mart, 1997). Much of the complexity in controlling any process comes from the complexity of the process being controlled. This complexity can be described in several ways. Highly non-linear systems are difficult to control, particularly when they have complex dynamics (such as instabilities to limit cycles and chaos). Difficulties can often be presented by constraints, either on the control parameters or in the operating regime. Lack of exact knowledge of the process, of course, makes control more difficult. Optimal control of many processes also requires systems, which make use of predictions of future behavior. The mathematical models for the dynamic systems are assumed to be non-linear differential equations. The goal of having these models is to capture the dynamics of non-linear processes, so as to have a way of controlling this dynamics for industrial purpose (Melin & Castillo, 1997).

We need a mathematical model of the non-linear dynamic system to understand the dynamics of the processes involved in the evolution of the system. For a specific case, this may require testing several models before obtaining the appropriate mathematical model for the process (Melin & Castillo, 1998). For real world systems with complex dynamics, we may even need several models for different sets of parameter values to represent all of the possible behaviors of the system. A general mathematical model of a dynamic system can be expressed as follows:

$$dx/dt = f_1(x, D, \alpha) - \beta f_2(x, D, \alpha) \qquad (11.1)$$
$$dp/dt = \beta f_2(x, D, \alpha)$$

where $x \in R^n$ is a vector of state variables, $p \in R^m$ is a vector of outputs, $\beta \in R$ is a constant measuring the efficiency of the conversion process, $D \in (0, 3)$ is the fractal dimension of the process, and $\alpha \in R$ is a selection parameter. The fractal dimension is used to characterize the process, for example in the case of biochemical reactors D represents the fractal dimension of the bacteria used for production (Castillo & Melin, 1994).

For a complex dynamical system it may be necessary to consider a set of mathematical models to represent adequately all of possible dynamic behaviors of the system. In this case, we need a decision scheme to select the appropriate model to use according to the linguistic value of a selection parameter α. We use a new fuzzy inference system for differential equations to achieve fuzzy modelling (Castillo & Melin, 1999). We have fuzzy rules of the form:

IF α is A_1 **AND** D is B_1 **THEN** M_1 (11.2)

... ...

IF α is A_n **AND** D is B_n **THEN** M_n

where $A_1, \ldots, A_n$ are linguistic values for α, $B_1, \ldots, B_n$ are linguistic values for the fractal dimension D, and $M_1, \ldots, M_n$ are mathematical models of the form given by equation (11.1). The selection parameter α can be the temperature for biochemical processes, because temperature changes cause the presence of new bacteria in this case (Melin & Castillo, 1998). For the case of aircraft dynamic systems, α can be related to environment parameters.

We combine adaptive model-based control using neural networks with the method for modelling using fuzzy logic and fractal theory, to obtain a new hybrid neuro-fuzzy-fractal method for control of non-linear dynamic systems. This general method combines the advantages of neural networks (ability for identification and control) with the advantages of fuzzy logic (ability for decision and use of expert knowledge) to achieve the goal of robust adaptive control of non-linear dynamic systems. We also use the fractal dimension to characterize the processes in modelling these dynamical systems. We have developed intelligent control systems using this new method for adaptive control for several applications, to validate our new approach for control. We have obtained very good results in controlling biochemical reactors and chemical reactors with the hybrid approach for control (Melin & Castillo, 1998). In this chapter, we describe the application of our new method to the case of controlling aircraft dynamic systems.

11.2 Fuzzy Modelling of Dynamical Systems

For a real-world dynamical system it may be necessary to consider a set of mathematical models to represent adequately all of the possible dynamic behaviors of the system. In this case, we need a fuzzy decision procedure to select the appropriate model to use according to the value of a selection parameter vector α. To implement this decision procedure, we need a fuzzy inference system that can use differential equations as consequents. For this purpose, we have developed a new fuzzy inference system that can be considered as a generalization of Sugeno's inference system (Sugeno & Kang, 1988), in which we are now using differential equations as consequents of the fuzzy rules, instead of simple polynomials like in the original Sugeno's method. Using this method, a fuzzy model for a general dynamical system can be expressed as follows:

$$\begin{array}{l}
\textbf{IF } \alpha_1 \text{ is } A_{11} \textbf{ AND } \alpha_2 \text{ is } A_{12} \ldots \textbf{ AND } \alpha_m \text{ is } A_{1m} \textbf{ THEN } dy/dt = f_1(y, \alpha) \\
\textbf{IF } \alpha_1 \text{ is } A_{21} \textbf{ AND } \alpha_2 \text{ is } A_{22} \ldots \textbf{ AND } \alpha_m \text{ is } A_{2m} \textbf{ THEN } dy/dt = f_2(y, \alpha) \\
\quad \vdots \qquad\qquad\qquad\qquad \vdots \\
\textbf{IF } \alpha_1 \text{ is } A_{n1} \textbf{ AND } \alpha_2 \text{ is } A_{n2} \ldots \textbf{ AND } \alpha_m \text{ is } A_{nm} \textbf{ THEN } dy/dt = f_n(y, \alpha)
\end{array} \tag{11.3}$$

where A_{ij} is the linguistic value of α_j for rule i-th, $\alpha \in R^m$ and is defined by $\alpha = [\alpha_1, \ldots, \alpha_m]$, and $y \in R^p$ is the output obtained by the numerical solution of the corresponding differential equation. Of course, it is assumed that each differential equation in (11.3) locally approximates the real dynamical system over a neighborhood (or region) of R^m.

The numerical solution of the differential equations can be achieved by the standard Runge-Kutta type method:

$$y_{n+1} = RK\,(y_n) + 1/2(k_1 + k_2) \qquad (11.4)$$
$$k_1 = hf(y_n, t_n)$$
$$k_2 = hf(y_n + k_1, t_{n+1})$$

where h is the step size of the method and RK can be considered as the Runge-Kutta operator that transforms numerical solutions from time n to time n+1. Numerical solutions are then aggregated by weighted average with weights obtained by the minimum of the firing strengths of the inputs:

$$y = \frac{w_1 y_1 + w_2 y_2 + \ldots + w_n y_n}{w_1 + w_2 + \ldots w_n} \qquad (11.5)$$

where:

$$y_1 = RK(f_1(y, \alpha))\; y_2 = RK(f_2(y, \alpha)) \ldots y_n = RK(f_n(y, \alpha))$$

The new fuzzy inference system for differential equations can be illustrated as in Figure 11.1, where a complex dynamical system is modeled by using four different mathematical models (M_1, M_2, M_3 and M_4). The decision scheme can be expressed as a single-input fuzzy model as follows:

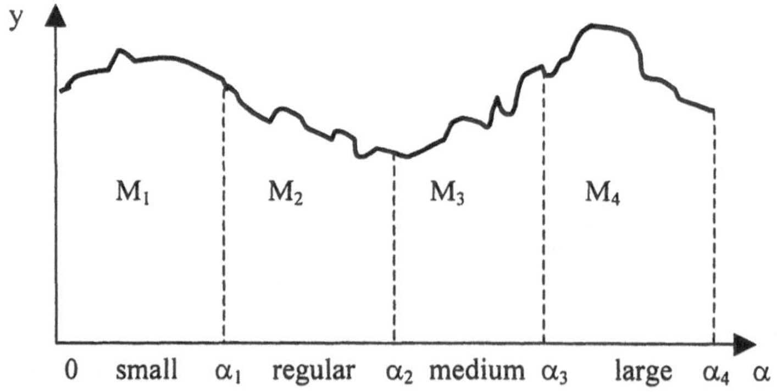

Figure 11.1 Modelling a complex dynamical system with the new fuzzy system.

$$\begin{array}{llllll} \textbf{IF} & \alpha & \text{is} & \text{small} & \textbf{THEN} & dy/dt = f_1(y, \alpha) \\ \textbf{IF} & \alpha & \text{is} & \text{regular} & \textbf{THEN} & dy/dt = f_2(y, \alpha) \\ \textbf{IF} & \alpha & \text{is} & \text{medium} & \textbf{THEN} & dy/dt = f_3(y, \alpha) \\ \textbf{IF} & \alpha & \text{is} & \text{large} & \textbf{THEN} & dy/dt = f_4(y, \alpha) \end{array} \tag{11.6}$$

where the output y is obtained by the numerical solution of the corresponding differential equation.

11.3 Neural Networks for Control

Parametric Adaptive Control is the problem of controlling the output of a system with a known structure but unknown parameters. These parameters can be considered as the elements of a vector p. If p is known, the parameter vector θ of a controller can be chosen as θ^* so that the plant together with the fixed controller behaves like a reference model described by a difference (or differential) equation with constant coefficients (Narendra & Annaswamy, 1989). If p is unknown, the vector $\theta(t)$ has to be adjusted on-line using all the available information concerning the system.

Two distinct approaches to the adaptive control of an unknown system are (i) direct control and (ii) indirect control. In direct control, the parameters of the controller are directly adjusted to reduce some norm of the output error. In indirect control, the parameters of the system are estimated as p(t) at any time instant and the parameter vector $\theta(t)$ of the controller is chosen assuming that p(t) represents the true value of the system parameter vector.

When indirect control is used to control a non-linear system, the plant is parameterized using a mathematical model of the general form described in Section 1 and the parameters of the model are updated using the identification error. The controller parameters in turn are adjusted by backpropagating the error (between the identified model and the reference model outputs) through the identified model. A block diagram of such an adaptive system is shown in Figure 11.2.

The overall structure of the adaptive system proposed in this paper to control a non-linear dynamical system is the same as shown in Figure 11.2 and is independent of the specific model used to identify the system. The delayed values of the system input and system output form the inputs to the neural network Nc which generates the feedback control signal to the system. The parameters of the Neural Network Ni are adjusted by backpropagating the identification error e_i while those of the Neural Network Nc are adjusted by backpropagating the control error (between the output of the reference model and the identification model) through the identification model.

The mathematical model for the non-linear dynamic system in the time domain is generated by the method of modelling (described in Section 11.2) using the real data that is measured on-line in the system. On the other hand, the fractal module is used to characterize the process and this information is used to specify the mathematical model in the time and space domain. This scheme enables the dynamic changes of models according to the changes of on-line process identification. Our new method for adaptive model-based control combining Neural Networks, Fuzzy Logic and Fractal Theory differs from our previous approach of considering only the use of neural networks and models (Melin & Castillo, 1997).

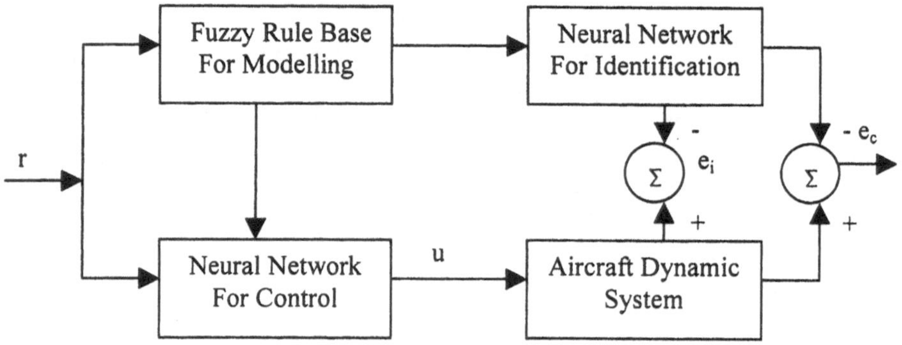

Figure 11.2. General Architecture for Adaptive Neuro-Fuzzy-Fractal Control.

11.4 Adaptive Control of Aircraft Systems

The mathematical models of aircraft systems can be represented as coupled non-linear differential equations (Melin & Castillo, 1998). In this case, we can develop a fuzzy rule base for modelling that enables the use of the appropriate mathematical model according to the changing conditions of the aircraft and its environment. For example, we can use the following model of an airplane when wind velocity is relatively small:

$$\begin{aligned} p' &= I_1(-q + l), \\ q' &= I_2(p + m) \end{aligned} \tag{11.7}$$

where I_1 and I_2 are the inertia moments of the airplane with respect to axis x and y, respectively, l and m are physical constants specific to the airplane, and p, q are the positions with respect to axis x and y, respectively. However, a more realistic model of an airplane in three dimensional space, is as follows:

$$\begin{aligned} p' &= I_1(-qr + l) \\ q' &= I_2(pr + m) \\ r' &= I_3(-pq + n) \end{aligned} \tag{11.8}$$

where now I_3 is the inertia moment of the airplane with respect to the z axis, n is a physical constant specific to the airplane, and r is the position along the z axis. Considering now wind disturbances in the model, we have the following equation:

$$\begin{aligned} p' &= I_1(-qr + l) - u_g \\ q' &= I_2(pr + m) \\ r' &= I_3(-pq + n) \end{aligned} \tag{11.9}$$

where u_g is the wind velocity. The magnitude of wind velocity is dependent on the altitude of the airplane in the following form:

$$u_g = u_{wind510} \left(1 + \frac{\ln(r/510)}{\ln(51)} \right) \tag{11.10}$$

where $u_{wind510}$ is the wind speed at 510 ft altitude (typical value = 20 ft/sec).

If we use the models of Equations (11.7)-(11.9) for describing aircraft dynamics, we can formulate a set of rules that relate the models to the conditions of the aircraft and its environment. Lets assume that M_1 is given by Equation (11.7), M_2 is given by Equation (11.8), and M_3 is given by Equation (11.9). Now using the wind velocity u_g and inertia moment I_1 as parameters, we can establish the fuzzy rule base for modelling as in Table 11.1.

In Table 11.1, we are assuming that the wind velocity u_g can have only two possible fuzzy values (small and large). This is sufficient to know if we have to use the mathematical model that takes into account the effect of wind (M_3) for u_g large or if we don't need to use it and simply the model M_2 is sufficient (for u_g small). Also, the inertia moment (I_1) helps in deciding between models M_1 and M_2 (or M_3).

Table 11.1. Fuzzy rule base for modelling aircraft dynamic systems.

IF			THEN
Wind Velocity	Inertia Moment	Fractal dimension	Model
Small	Small	Low	M_1
Small	Small	Medium	M_2
Small	Large	Low	M_2
Small	Large	Medium	M_2
Large	Small	Medium	M_3
Large	Large	Medium	M_3
Large	Large	High	M_3

11.5 Experimental Results

To give an idea of the performance of our neuro-fuzzy-fractal approach for adaptive control, we show below simulation results for aircraft dynamic systems. First, we show in Figure 11.3 the fuzzy rule base for a prototype intelligent system developed in the fuzzy logic toolbox of the MATLAB programming language. We show in Figure 11.4 the non-linear surface for the problem of aircraft dynamics using as input variables: fractal dimension and wind velocity.

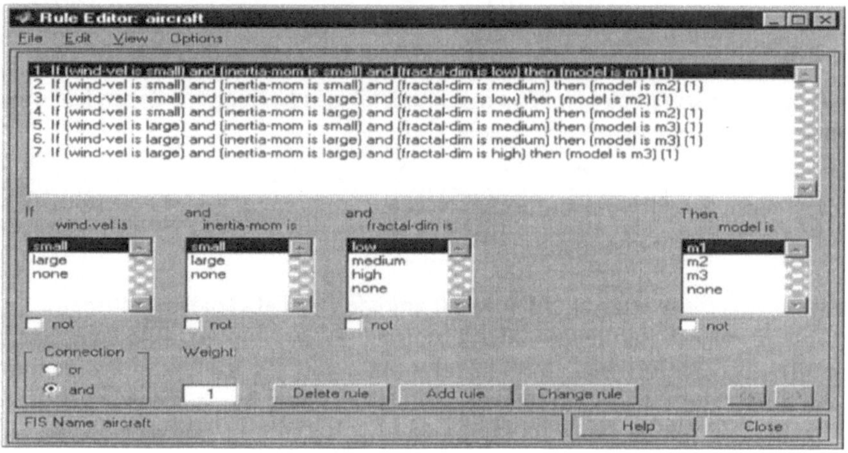

Figure 11.3. Fuzzy rule base

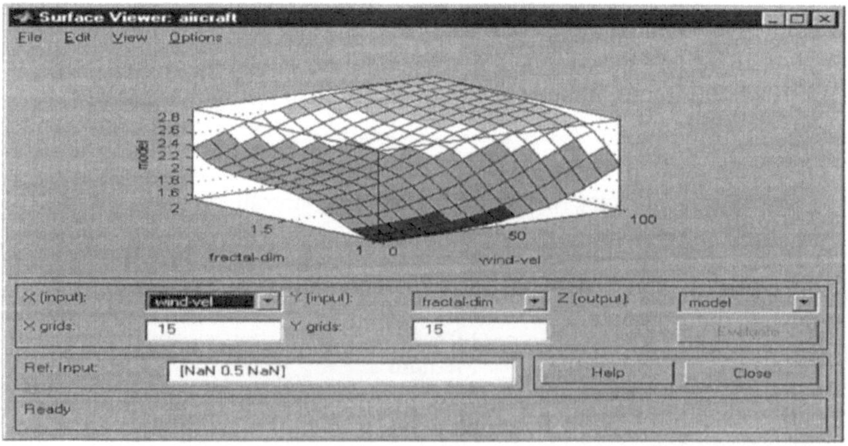

Figure 11.4 Non-linear surface for aircraft dynamics

We show simulation results for an aircraft system obtained using our new method for modelling dynamical systems. In Figure 11.5 and Figure 11.6 we show results for an airplane with inertia moments: $I_1 = 1$, $I_2 = 0.4$, $I_3 = 0.05$ and the constants are: $l = m = n = 1$. The initial conditions are: $p(0) = 0$, $q(0) = 0$, $r(0) = 0$.

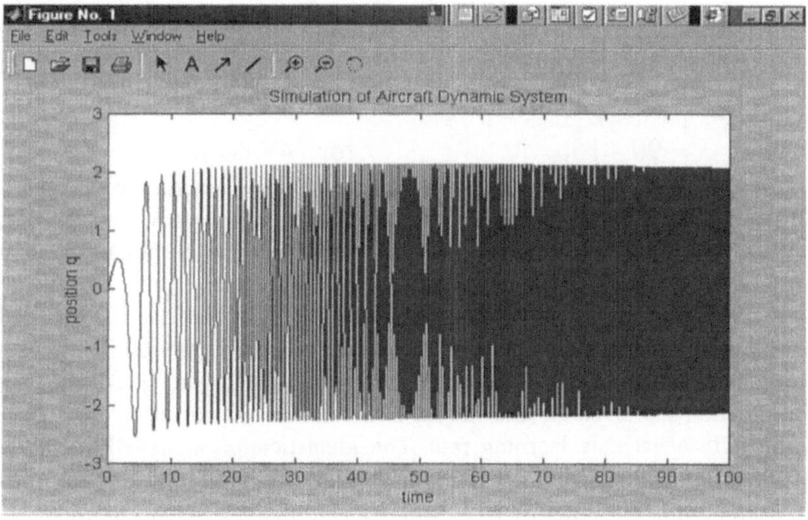

Figure 11.5 Simulation of position q

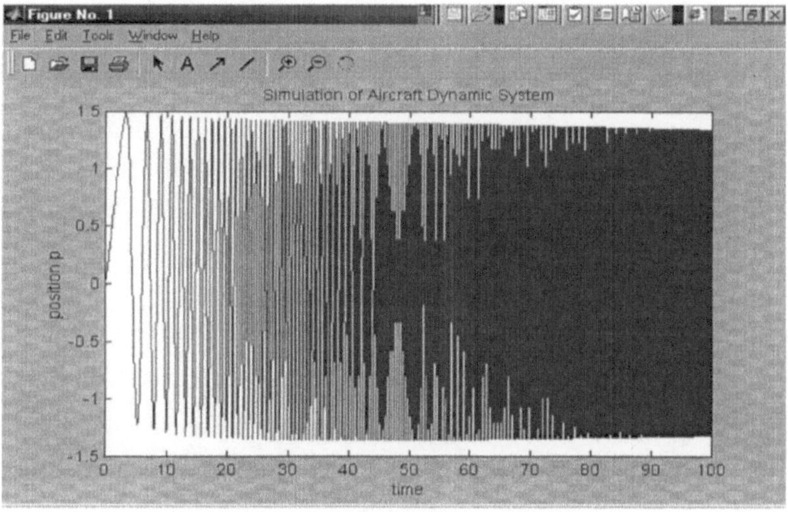

Figure 11.6 Simulation of position p

To give an idea of the performance of our neuro-fuzzy approach for adaptive model-based control of aircraft dynamics, we show below (Figure 11.7) simulation results obtained for the case of controlling the altitude of an airplane for a flight of 6 hours. We assume that the airplane takes about one hour to achieve the cruising altitude 30 000 ft, then cruises along for about three hours at this altitude (with minor fluctuations), and finally descends for about two hours to its final landing point. We will consider the desired trajectory as follows:

$$r_d = \begin{cases} 30t + \sin 2t & \text{for } 0 \leq t \leq 1 \\ 30 + 2\sin 10t & \text{for } 1 < t \leq 4 \\ 90 - 15t & \text{for } 4 < t \leq 6 \end{cases}$$

Of course, a complete desired trajectory for the airplane would have to include the positions for the airplane in the x and y directions (variables p, q in the models). However, we think that here for illustration purposes is sufficient to show the control of the altitude r for the airplane.

We used three-layer neural networks (with 10 hidden neurons) with the Levenberg-Marquardt algorithm and hyperbolic tangent sigmoidal functions as the activation functions for the neurons. We show in Figure 11.7 the function approximation achieved by the neural network for control after 800 epochs of training with a variable learning rate. The identification achieved by the neural network (after 800 epochs) can be considered very good because the error has been decreased to the order of 10^{-1}. Still, we can obtain a better approximation by using more hidden neurons or more layers. In any case, we can see clearly (from Figure 11.7) how the neural network learns to control the aircraft, because it is able to follow the arbitrary desired trajectory.

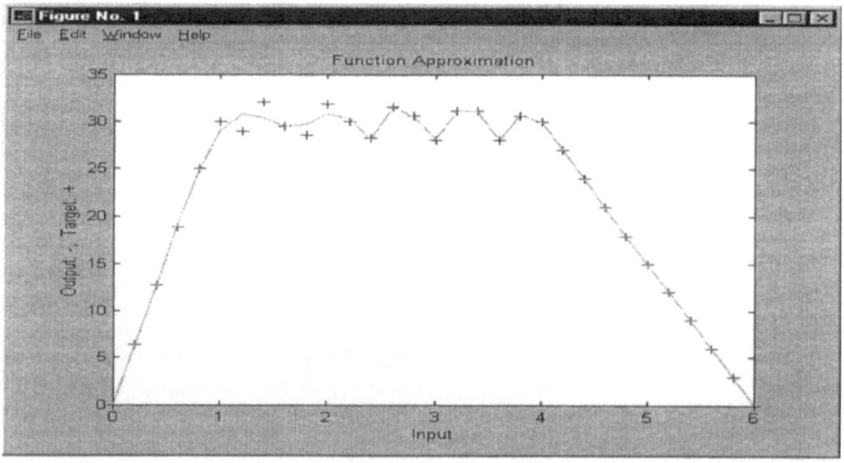

Figure 11.7 Function approximation of the neural network for control of an airplane.

We have to mention here that these simulation experiments for the case of a specific flight for a given airplane show very good results. We have also tried our approach for control with other types of flights and airplanes with very good results. Of course, experimentation with real aircraft dynamic systems is still needed to really measure the power of the intelligent control approach. We expect that the efficiency of our method will be reduced slightly by real disturbances when applied to a real aircraft system. However, no major changes in hybrid architecture for the intelligent control system are expected.

11.6 Summary

We have developed a general method for adaptive model based control of non-linear dynamic systems using Neural Networks, Fuzzy Logic and Fractal Theory. We illustrated our hybrid method for control with the case of controlling aircraft dynamics. In this case, the mathematical models represent the aircraft dynamics during flight. We also described in this chapter an intelligent adaptive controller based on the use of neural networks, fuzzy logic and mathematical models for the system. The proposed adaptive controller performs rather well considering the complexity of the domain being considered in this research work. We have shown that our method can be used to control chaotic and unstable behavior in aircraft systems. Chaotic behavior has been associated with the "flutter" effect in real airplanes, and for this reason is very important to avoid this kind of unstable behavior. We can say that combining Neural Networks, Fuzzy Logic and Fractal Theory, using the advantages that each of these methodologies has, can give good results for this kind of application. Also, we believe that our neuro-fuzzy-fractal approach is a good alternative for solving similar problems. Finally, we have to mention here that genetic algorithms could be used to increase even more the efficiency of the hybrid system for control. Genetic algorithms may be used to optimize the membership functions needed in the fuzzy rule base for modelling, or to optimize the architecture of the neural networks for control and identification.

Chapter 12

Controlling Electrochemical Processes

We describe in this chapter, different hybrid approaches for controlling dynamical systems in manufacturing applications. The hybrid approaches combine soft computing techniques and mathematical models to achieve the goal of controlling the manufacturing process to follow a desired production plan. We have developed several hybrid architectures that combine fuzzy logic, neural networks, and genetic algorithms, to compare the performance of each of these combinations and decide on the best one for our purpose. Electrochemical processes, like the ones used in battery formation, are very complex and for this reason very difficult to control. We have achieved very good results using fuzzy logic for control, neural networks for modelling the process, and genetic algorithms for tuning the hybrid intelligent system. For this reason, we consider that the neuro-fuzzy-genetic approach is the most appropriate for this case.

12.1 Introduction

The dynamics of an electrochemical system is non-linear and for this reason can be very difficult to predict in an accurate manner (Melin & Castillo, 1999). Also, mathematical models of electrochemical processes are difficult to derive and they are not very accurate. Traditionally, the models that have been used for electrochemical processes are from Statistics, but these models do not approximate the dynamic behavior of the processes with the accuracy required in practice. We need adaptive control of the electrochemical process to achieve on-line control of the production line. Of course, adaptive control is easier to achieve if one uses a reference model of the process (Melin & Castillo, 1998). The problem is how to obtain a good model for the process, considering that mathematical models are very inefficient. The use of soft computing techniques has been recognized as a different approach to modeling with good results, and for this reason we decided to turn to these techniques for our problem (Castillo &

Melin, 1999). We have tested fuzzy logic and neural networks for modeling the electrochemical processes and we have decided that the best choice for this case is to use neural networks. The other part of the problem is how to control the non-linear electrochemical process in the desired way to achieve the production with the required quality. We tested different techniques and we arrived to the conclusion that fuzzy logic (Zadeh, 1975) was the best choice for this case. We developed a set of fuzzy rules with the expert knowledge for controlling the manufacturing process. The membership functions for the linguistic variables in the rules were tuned using a specific genetic algorithm. The genetic algorithm was used for searching the parameter space of the membership functions using real data from production lines. Our particular neuro-fuzzy-genetic approach has been implemented as an intelligent system to control the formation of batteries in a real plant with good results. We think that this hybrid approach can be used for controlling similar processes, the only change that will be required is to adapt the parameters and rules to the new application.

12.2 Problem Description

In a battery a process of conversion of chemical energy into electrical energy is carried out. The chemical energy contained in the electrode and electrolyte is converted into electrical power by means of electrochemical reactions. When connecting the battery to a source of direct current a flow of electrons takes place for the external circuit, and of ions inside the battery, giving an accumulation of load in the battery. The quantity of electric current that is required to load the battery is determined by an unalterable law of nature, that was postulated by Michael Faraday, which is known as the Law of Faraday (Bode, Brodd & Kordesch, 1977). Faraday found that the quantity of electric power required to perform an electrochemical change in a metal is related to the relative weight of the metal. In the specific case of lead this is considered to be 118 amperes hour for pound of positive active material for cell. In practice, more energy is required to counteract the losses due to the heat and to the generation of gas.

We show in Table 12.1 experimental data for a specific type of battery with different sizes of the plates, and different number of plates for each cell. In this table, we show the load time and the average current needed for the respective load. In Table 12.1 we can observe that to form a battery we need to apply a particular current intensity during a certain amount of time to achieve the required loading for the battery. For example, from the first line of Table 12.1 we can see that a battery with positive plate of 0.060" and negative plate of 0.050" requires 155 amperes in 1 hour, or 2.2 amperes during 72 hours.

The goal of the manufacturers of batteries is to reduce the time required to load the battery. However, current intensity can't be increased arbitrarily because of the physical characteristics of the specific battery. If the current is increased too much, the temperature in the battery will go over a safe temperature value eventually causing the destruction of the battery. The problem is then, of finding out how much can we increase the current without causing the battery to explode. Of course, we need to control the current during the process of battery formation, and we want to do this in such a way as minimize the time of loading.

With the purpose of finding out, during the process of formation of the battery, the appropriate values of current intensity without surpassing the limits of temperature (Hehner & Orsino, 1985), we propose three systems for intelligent control of the process. The first one uses only fuzzy logic for control and statistical models of the process. The second one uses a neuro-fuzzy approach to develop the fuzzy rules controlling the electrochemical process. The third approach uses neural networks to model the process, fuzzy rules for control and genetic algorithms to tune the membership functions. Of course, this last approach gave us better results as we will show later in the chapter.

Table 12.1 Experimental data for different types of batteries

Type of Plate						
	Positive 0.060" Negative 0.050"			Positive 0.070" Negative 0.060"		
Plate cell	Total A. H.	72 hr Amp.	96 hr Amp.	Total A.H.	72 hr Amp	96 hr Amp
7	155	2.2	1.6	165	2.4	1.8
9	180	2.8	2.0	200	2.8	2.2
11	2230	3.2	2.4	245	3.4	2.4
13	260	3.6	2.6	295	4.0	3.0
15	300	4.2	3.0	345	4.8	3.6
17	400	5.6	4.2	415	5.8	4.4

12.3 Fuzzy Method for Control

In this approach we use a statistical model to represent the electrochemical process and a fuzzy rule base for process control. The temperature in the battery depends on the electrical current that circulates in it during its formation, this means that to maintain the temperature below a specific threshold it is important to control the intensity of the current. Therefore for this case the independent variable is the average current I, and the dependent variable is the average temperature T. A simple statistical model can stated as follows:

$$T = \beta_o + \beta_1 I \qquad (12.1)$$

where β_o and β_1 are parameters to be estimated (by least squares) using real data for this problem. In Table 12.2, we show experimental values for a battery of 6

Volts, which according to manufacturer's specifications should be loaded by using 200 amperes hour.

Using the data from Table 12.2 we can obtain (by least squares method) the values of β_o and β_1 (Sepulveda, Castillo, Montiel, & Lopez, 1998). The equation is as follows:

$$T = 88.03 + 2.5304\ I \qquad (12.2)$$

with correlation value of only 0.57 which is understandable because of the complexity of the data.

For the fuzzy controller we used as input variables, the temperature T and the change of temperature dT/dt, and as output variable the current intensity that should be applied to the battery. In Figure 12.1 we show the architecture of our control system.

Table 12.2 Measured values of temperature and electrical current for a battery of 200 amperes hour.

Hrs	T	I	Hrs	T	I
21:00	111	5.22	23:00	93	3.53
23:00	100	5.21	1:00	91	3.40
1:00	105	5.52	3:00	92	3.32
3:00	100	5.66	5:00	96	3.16
5:00	100	5.60	7:00	98	3.10
7:00	97	5.72	9:00	98	3.14
9:00	92	4.82	11:00	102	3.12
11:00	95	4.32	13:00	99	3.03
13:00	102	4.10	15:00	98	3.05
15:00	103	4.05	17:00	97	3.06
17:00	100	3.40	19:00	95	2.96
19:00	97	3.77	21:00	94	2.60
21:00	94	3.62	23:00	96	2.76

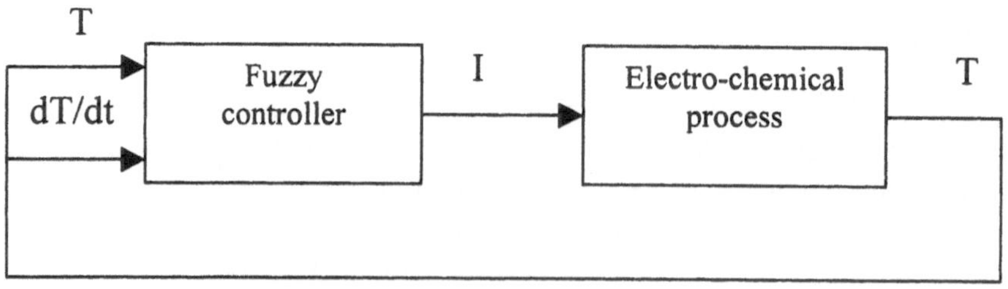

Figure 12.1 Fuzzy Control of the process

The control method was implemented in the MATLAB programming language. For each of the linguistic variables in the fuzzy system it was considered convenient to use five linguistic terms. In Figure 12.2 we show the fuzzy rule base implemented in the Fuzzy Logic Toolbox of MATLAB. We have 25 rules because we are using 5 linguistic terms for each variable. The membership functions were tuned manually until they give the best values for the problem. We discuss the results of this approach later in the chapter.

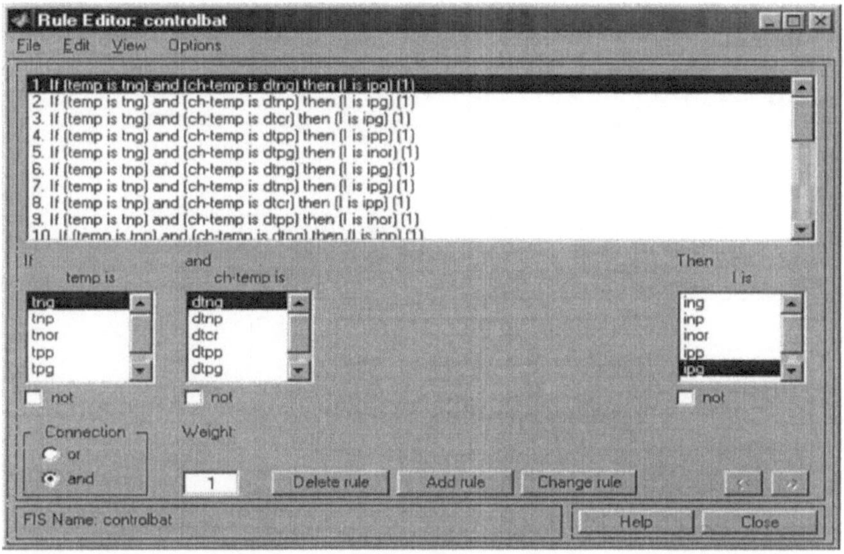

Figure 12.2 Fuzzy rule base for controlling the Process

12.4 Neuro-Fuzzy Method for Control

Since it is difficult to tune a particular inference system to model a complex dynamical system (Castillo & Melin, 1998) it is convenient to use adaptive fuzzy inference systems. Adaptive neuro-fuzzy inference systems (ANFIS) can be used to adapt the membership functions and consequents of the rule base according to historical data of the problem (Jang, Sun & Mizutani, 1997). The ANFIS methodology has been applied to many real world problems with good results, and we consider it as a good option to investigate for this application. In this case, we can use the data from Table 12.2 and apply the ANFIS methodology to find the best fuzzy system for our problem. We used the fuzzy logic toolbox of MATLAB to apply the ANFIS methodology to our problem with 5 membership functions and first order Sugeno functions in the consequents.

In Figure 12.3 we show the architecture of the intelligent control system. The fuzzy rule base was implemented in the MATLAB programming language. We show in Figure 12.4 the non-linear surface for control. We only used 5 rules because there is only one input variable (temperature) and one output variable (electrical current) with 5 linguistic variables each. We describe in detail the results of this approach later in the chapter, but we can anticipate better results that in the previous approach using only fuzzy logic.

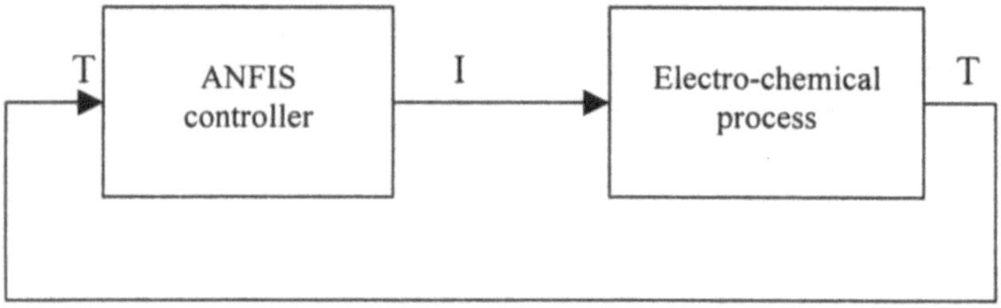

Figure 12.3 ANFIS Control of the process

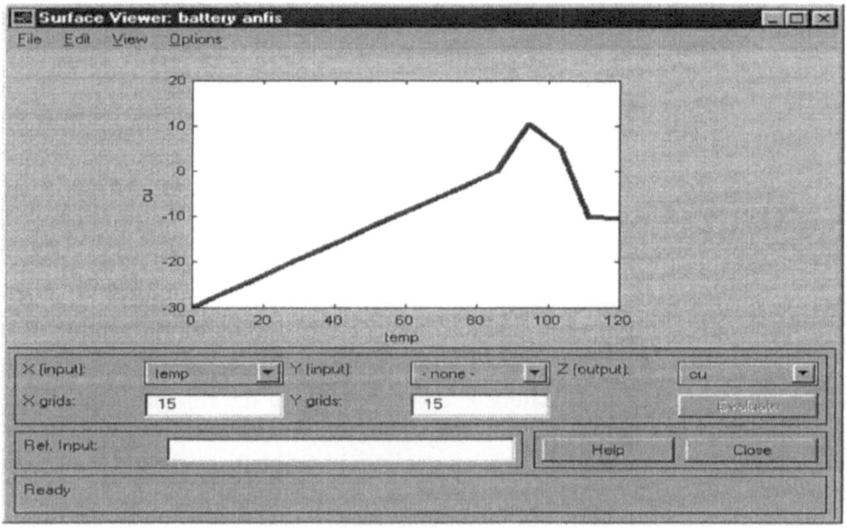

Figure 12.4 ANFIS surface for the process.

12.5 Neuro-Fuzzy-Genetic Method for Control

In this case, we consider using neural networks for modelling the electrochemical process, fuzzy logic for controlling the electrical current in the process, and genetic algorithms for adapting the membership functions of the fuzzy system (Castillo & Melin, 1998). The general architecture of the intelligent control system is shown in Figure 12.5. In this figure, we can see clearly the basic idea of the hybrid approach, which is to use the advantages of each SC technique to achieve efficient real time control.

A multilayer feedforward neural network was used for modelling the electrochemical process. We used the data form Table 12.2 and the Levenberg-Marquardt learning algorithm to train the neural network. We used a three layer neural network with 15 nodes in the hidden layer. The results of training for 2000 epochs are shown in Figure 12.6. The sum of squared errors was reduced from about 200 initially to 11.25 at the end, which is a very good approximation in this case due to the complexity of the problem.

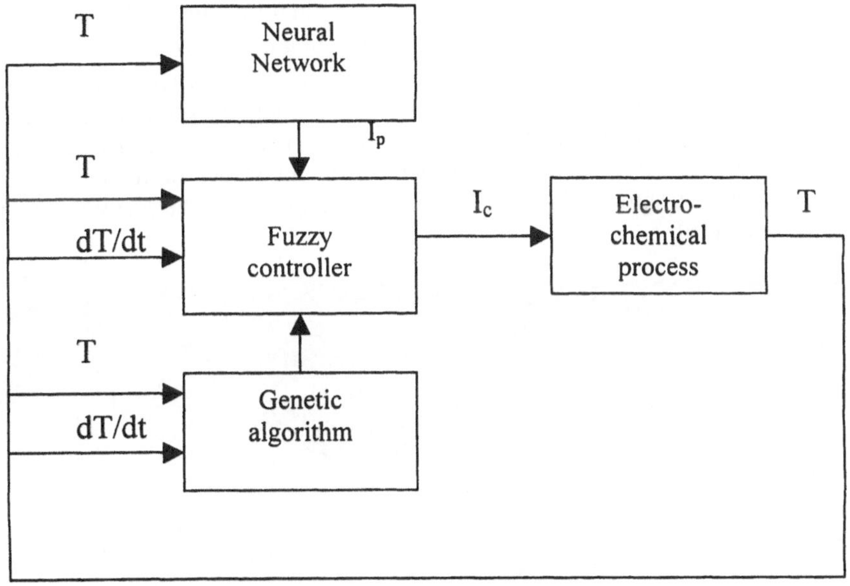

Figure 12.5 Neuro-Fuzzy-Genetic Control

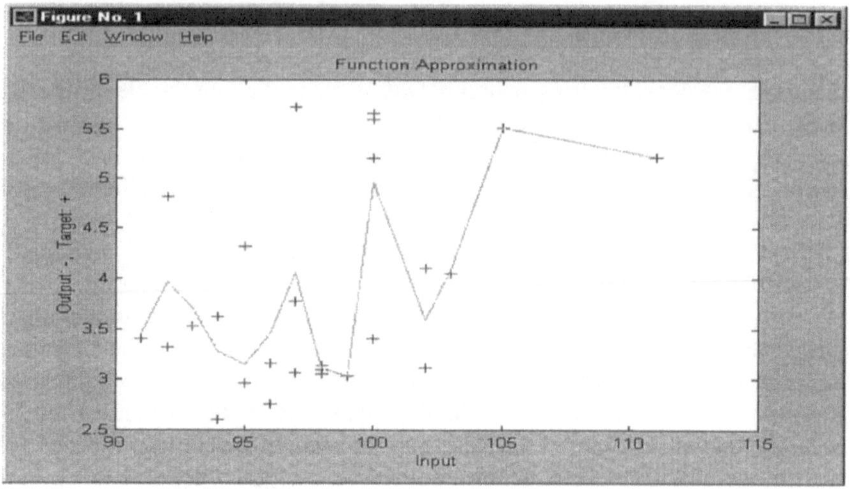

Figure 12.6 Final training for the neural network.

The fuzzy rule base was implemented in the Fuzzy Logic Toolbox of MATLAB. We used 25 fuzzy rules because there were 5 linguistic terms for each input variable. The initial membership functions were obtained from experts, but they were improved using a specific genetic algorithm to arrive to the final membership functions. The genetic algorithm had mutation, single crossover and elitism. The mutation rate was 0.05, and the crossover rate was 0.1. These specific values were found by experimentation with different values until the best ones were selected. We show in Figure 12.7 the implementation of the fuzzy system for intelligent control.

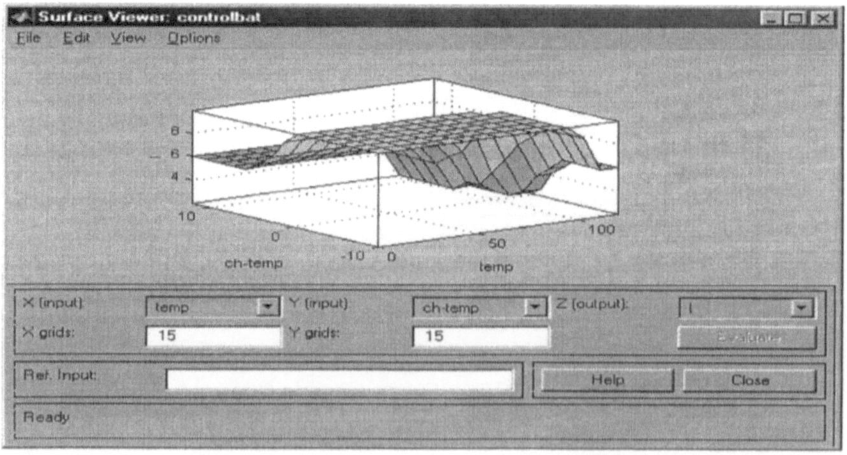

Figure 12.7 Non-linear surface for control.

12.6 Experimental Results for the Three Hybrid Approaches

We compared the three hybrid intelligent control systems by simulating the formation (loading) of a 6 Volts battery. This particular battery is manually loaded (in the real manufacturing plant) by applying 2 amperes for 50 hours under manufacturer's specifications. We show in Table 12.3 the results of the three hybrid methods for control.

Table 12.3 Comparison of the methods for control

Control Method	Time for Loading
Manual Control	50 hours
Conventional Control	36 hours
Fuzzy Control	32 hours
Neuro-Fuzzy Control	30 hours
Neuro-Fuzzy-Genetic	25 hours

We can see from Table 12.3 that the fuzzy control method reduces 36% the time required to form the battery compared with manual control, and 11.11% compared with conventional PID control (Sepulveda, Castillo, Montiel, Ross, 1998). We can also see how ANFIS helps in reducing even more this time of formation because we are using neural networks for adapting the system. Now the reduction is of 40% with respect to manual control. Finally, we can notice that using a neuro-fuzzy-genetic approach reduces even more the time of formation because the genetic algorithm optimizes even more the fuzzy system for control and we are also using the neural network as a model of the process. In this case, the reduction is of 50 % with respect to manual control. These results are only for a specific type of battery (the 6 Volts type of battery). For other types of batteries the results are similar, with the best results for the neuro-fuzzy-genetic approach, but reductions in time oscillate between 40% and 50%. From these results, we can see very clearly that combining different soft computing techniques in the right way, we can increase performance of an intelligent system for control. Of course, we are not suggesting that we always need to combine techniques, in some cases this may not be necessary. Also, in other type of applications the best hybrid combination needed may not be the neuro-fuzzy-genetic one used here, other combinations may be better for other situations.

12.7 Summary

We have described in this chapter, three different approaches for controlling a complex electrochemical process. We have implemented these hybrid approaches as intelligent systems for control in the MATLAB programming language. We compared the results of these systems with conventional PID control, to measure the efficiency of the hybrid intelligent control systems. We have shown that for this type of application the use of several, soft computing, techniques can help in reducing the time required to produce a battery in the manufacturing plant. Even fuzzy control alone can reduce the formation time, but using neural networks and genetic algorithms reduces even more the time for production. Of course, this means that manufacturers can produce the batteries in roughly half the time needed before. These are very good results for this application. A hardware implementation is still needed to completely automate the control of the process.

Chapter 13

Controlling International Trade Dynamics

We describe in this chapter the application of our new method for adaptive model-based control (using a neuro-fuzzy-genetic approach) to the problem of controlling international trade dynamics. The problem of international trade between two or more countries is a very complex one because of the non-linearities involved in the mathematical models (Castillo and Melin, 1998). In this chapter, we describe the methodology to develop an intelligent system for controlling international trade that can be used by the government of a specific country to maximize the profit from its international trade with other countries. Our method for adaptive model-based control of non-linear dynamical systems consists of using a fuzzy rule base for model selection, a genetic algorithm for identification and a neural network for control (Melin and Castillo, 1998). Accordingly, an intelligent control system based on our methodology has an architecture with three main modules: model selection, identification and control. For the case of international trade, we have developed the fuzzy rule base for selecting the appropriate mathematical models for the problem, the genetic algorithm for parameter identification, and the neural network for control.

13.1 Introduction

Mathematical modelling of international trade has been done traditionally with linear statistical models from classical Econometric Theory (Gujarati, 1987). However, more recently some researchers have found statistical evidence (Anderson, Arrow, and Pines, 1988) that time series from financial and economical variables show erratic fluctuations in time. It is well known that simple linear models can not represent this erratic dynamic behavior that has been found in several time series. Then, it becomes necessary to use non-linear mathematical models that will enable us to represent this complex dynamic behavior found for systems in economics or finance. Non-linear models from the theory of Dynamical Systems can show the behavior known as "chaos" for different ranges of parameter values (Devaney, 1989) and for this reason they become a good choice in modelling complex financial or economic problems. Our experience on this matter (Castillo and Melin, 1995a, 1995b, 1996, 1997) has shown that this line of research is very promising.

The mathematical models of international trade can be represented as systems of coupled non-linear differential equations. The general mathematical model for describing the international trade between two or more countries can be written as follows:

$$Y'_i = \alpha_i(I_i - S_i + \gamma\,(EX_i - IM_i)) \qquad i = 2, 3, \ldots$$
$$r'_i = \beta_i\,(L_i - M_i/P_i)$$

where: Y_i is the national income, r_i is the interest rate, M_i is the nominal money supply, P_i are the goods prices, I_i national gross investment, S_i national savings, EX_i is the total exports, IM_i is the total imports and α_i, β_i are parameters to be estimated. We use parameter γ as a decision parameter in the fuzzy rule base for model selection, because we consider that γ measures the effect of international trade in the economies. We can then formulate a fuzzy rule base for model selection (based on the selection parameter γ) that enables the use of the appropriate specific mathematical model according to the changing conditions of the economies involved.

We use a specific genetic algorithm to estimate the parameters of the specific mathematical model of international trade being considered at the moment by the fuzzy selection process. Parameter estimation for the models (identification) is performed by using a genetic algorithm for the minimization of an objective function measuring how well the specific model is fitting real data for the problem (time series).

We use a feedforward multilayer neural network for control with the Levenberg-Marquardt learning algorithm (Jang, Sun and Mizutani, 1997). We have obtained simulation results for the case of three countries (USA, Canada and Mexico) with international trade.

13.2 Mathematical Modelling of International Trade

Mathematical modelling of problems in finance and economics has been always very important in the applications, since it enables the simulation and forecasting of the relevant financial and economical variables. Mathematical modelling, from the theoretical point of view, contributes to the understanding of economical and financial phenomena, since it enables the investigation of possible relationships between different financial and economical variables. On the other hand, mathematical modelling, from the practical point of view, can be used as a tool for planning and decision making in the industry or in government.

Mathematical modelling of international trade has been done traditionally with linear statistical models from classical Econometric Theory (Gujarati, 1987). However, more recently some researchers have found statistical evidence (Anderson, Arrow, and Pines, 1988) that time series from financial and economical variables show erratic fluctuations in time. It is well known that simple linear models can not represent this erratic dynamic behavior that has been found in several time series. Then, it becomes necessary to use non-linear mathematical models that will enable us to represent this complex dynamic behavior found for systems in economics or finance. Non-linear models from the theory of Dynamical Systems can show the behavior known as "chaos" for different ranges of parameter values (Devaney, 1989, Lorenz, 1987) and for this reason they become a good choice in modelling complex financial or economic problems. Our experience on this matter has shown that this line of research is very promising.

We will consider first modelling the dynamics of autonomous economies, i.e.., study the oscillations of an autonomous economy. Then, we will consider modelling the problem of International Trade as a perturbation of the internal oscillations of an autonomous economy.

13.2.1 Oscillations in Autonomous Economies

Consider the Keynesian macroeconomic model (Lorenz, 1987) of a single economy with Y as income, r as the interest rate, M as the (constant) nominal money supply, and assume that the good prices, P, are fixed during the relevant time interval. Suppose that gross investment, I, and savings, S, depend both on income and the interest rate in the familiar way, i.e.,

$$I = I\,(Y,r) \quad , \qquad Iy > 0,\ Ir < 0$$
$$S = S\,(Y,r) \quad , \qquad Sy > 0,\ Sr < 0$$

Income adjusts according to excess demand in the goods market, i.e.,

$$Y' = \alpha\,(I - S) \qquad \alpha > 0 \tag{13.1}$$

The set of points $\{(Y,r)|\ I(Y,r) = S(Y,r)\}$ constitutes the IS-curve of the model. Let L(Y,r) denote the liquidity preference with $Ly > 0$, $Lr < 0$ and assume that the interest rate adjusts according to:

$$r' = \beta\,(L(Y,r) - M/p) \quad , \beta > 0 \tag{13.2}$$

with the set of points $\{(Y,r)|\ L(Y,r) = M/p\}$ forming the LM-curve of the model. As is well known, the equilibrium (Y^*,r^*) is asymptotically (locally) stable if trJ < 0 and det J >0, where J is the Jacobian of the system and Tr=trace, det = determinant. Also, it can be demonstrated by means of the Poncaré-Bendixon Theorem that system (13.1) - (13.2) is able to generate oscillating behavior.

13.2.2 International Trade as a Perturbation of Internal Oscillations

Consider three economies, each of which is described by equations like (13.1)-(13.2) with possibly different numerical specifications of the functions, i.e.,

$$\begin{aligned} Y'i &= \alpha\, i\,(Ii(Yi,ri) - Si(Yi,ri)) \\ & \qquad\qquad\qquad\qquad i = 1, 2, 3 \\ r'i &= \beta\, i\,(Li(Yi,ri) - Mi/pi) \end{aligned} \tag{13.3}$$

Equation (13.3) constitutes a six-dimensional differential equation system, which can also be written as a system of three independent two-dimensional limit cycle oscillators.

By introducing international trade with linear functions $EXi=EXi(Yj,Yk)$, $i\neq j,k$ and $Imi=Imi(Yi)$, equation (13.3) becomes:

$$\begin{aligned} Y'i &= \alpha i\,(Ii(Yi,ri) - Si(Yi,ri) + EXi(Yj,Yk) - Imi(Yi)) \\ r'i &= \beta i\,(Li(Yi,ri) - Mi/pi) \end{aligned} \tag{13.4}$$

with i, j, k= 1, 2, 3; j, k = i, and Mi as the money supplies reflecting balance of payments equilibria. Equation (13.4) constitutes a system of three linearly coupled limit cycle oscillators. The following theorem can then be demonstrated for system (13.4).

Theorem 13.1: If all three autonomous economies are oscillating, the introduction of international trade may imply the existence of a strange attractor (chaotic behavior).

Of course, chaotic behavior may occur for certain ranges of parameter values for the αi, βi, Mi parameters. However, the emergence of strange attractors is not exclusive in models like these: some variations in the parameters can lead to the occurrence of other phenomena like quasi-periodic motion or phase-locking. The main goal for a certain country is to achieve a stable behavior in its economy while in this International Trade System, in this way controlling its future behavior in this system. As a result of this, a specific country (like Mexico) can optimize its profit while in a system of three countries (like with the NAFTA trade agreement).

13.3 Fuzzy Logic for Model Selection

For a complex dynamical system it may be necessary to consider a set of mathematical models to represent adequately all of the possible dynamic behaviors of the system (Melin & Castillo, 1998). In this case, we need a decision scheme to select the appropriate model to use according to the value of a selection parameter α. In this section, we show a method for model selection based on fuzzy logic and a new fuzzy inference system for differential equations.

We have designed a method, based on fuzzy logic techniques, for mathematical model selection using as input the numerical value of a selection parameter α. We assume, in what follows, that parameter α is defined over a real-valued interval:

$$\alpha_0 \leq \alpha \leq \alpha_n . \tag{13.5}$$

We also assume that we have n mathematical models considered appropriate for the respective n subintervals, defined on $[\alpha_0 , \alpha_n]$, as follows:

$$\alpha_0 \leq \alpha < \alpha_1 , \quad \alpha_1 \leq \alpha < \alpha_2 ,..., \qquad \alpha_{n-1} \leq \alpha \leq \alpha_n . \tag{13.6}$$

The corresponding n mathematical models for these subintervals can be expressed as differential equations:

$$\begin{aligned} dy/dt &= f_1(y, \alpha) \\ dy/dt &= f_2(y, \alpha) \\ &\dots \\ dy/dt &= f_n(y, \alpha). \end{aligned} \tag{13.7}$$

Then, we can define a set of fuzzy if-then rules that basically relate the subintervals to the mathematical models in a one-to-one fashion. The advantage of using fuzzy rules (instead of conventional simple if-then rules) is that we can manage the underlying uncertainty of this process of model selection. We show the set of fuzzy rules for model selection in Table 13.1.

Table 13.1 Decision scheme for model selection

IF	THEN	
$\alpha_0 \leq \alpha < \alpha_1$	M_1:	$dy/dt = f_1(y, \alpha)$
$\alpha_1 \leq \alpha < \alpha_2$	M_2:	$dy/dt = f_2(y, \alpha)$
$\alpha_2 \leq \alpha < \alpha_3$	M_3:	$dy/dt = f_3(y, \alpha)$
$\vdots$		$\vdots$
$\alpha_{n-1} \leq \alpha \leq \alpha_n$	M_n:	$dy/dt = f_n(y, \alpha)$

To implement this decision scheme, we need a reasoning method that can use differential equations as consequents. We have developed a new fuzzy inference system that can be considered as a generalization of Sugeno's inference system (Sugeno & Kang, 1988) in which we are now considering differential equations as consequents of the fuzzy rules, instead of simple polynomials. Using this method, the decision scheme of Table 13.1 can be expressed as a single-input fuzzy model as follows:

$$\begin{cases} \text{If } \alpha \text{ is small} & \text{then } dy/dt = f_1(y,\alpha) \\ \text{If } \alpha \text{ is regular} & \text{then } dy/dt = f_2(y,\alpha) \\ \text{If } \alpha \text{ is medium} & \text{then } dy/dt = f_3(y,\alpha) \\ \quad\vdots & \quad\vdots \\ \text{If } \alpha \text{ is large} & \text{then } dy/dt = f_n(y,\alpha) \end{cases}$$

where the output y is obtained by the numerical solution of the corresponding differential equation. We have to note here that this new fuzzy inference system reduces to the standard Sugeno system only when the differential equations have closed-form solutions in the form of polynomials. However, the solutions to the differential equations can be more complicated analytical functions or in most cases the solutions are so complex that can only be approximated by numerical methods. The advantage of this generalization of Sugeno's original method is that, in general, we can represent more complicated dynamic behaviors and also because of this fact, the number of rules needed to represent a given dynamical system is smaller.

The reasoning procedure is very similar to the original Sugeno's procedure, except that now in the output we obtain the crisp values of "y" by solving numerically the corresponding differential equations. The numerical solutions of the differential equations can be achieved by the standard Runge-Kutta type method (Nakamura, 1997):

$$y_{n+1} = RK(y_n) = y_n + 1/2(k_1 + k_2)$$
$$k_1 = hf(y_n, t_n)$$
$$k_2 = hf(y_n + k_1, t_{n+1})$$

where h is the step size of the numerical method and RK can be considered as the Runge-Kutta operator that transforms numerical solutions from time n to time n+1.

The reasoning procedure for differential equations can also be used for rules with multiple inputs (for the case of several selection parameters) by simply considering the minimum of the firing strengths of each of the inputs. The fuzzy inference system for differential equations can be illustrated as in Figure 13.1, where a complex dynamical system is modeled by using four different mathematical models (M_1, M_2, M_3 and M_4).

Of course, for this decision scheme to work we need to define membership functions for the different values of the parameter α corresponding to the mathematical models. The membership functions for the models should give us the degree of belief that a particular model is the correct one for a specific value of the parameter α.

To apply this method of model selection, to a particular application, we have to find the corresponding selection parameter α (or even several parameters) to be used in the decision scheme proposed in Table 13.1. Then, a partition of the definition interval for α has to be performed. After this, the one-to-one map between the mathematical models and the subintervals (obtained from the partition) is constructed. In this way, we can obtain the fuzzy rule base for model selection for a particular application.

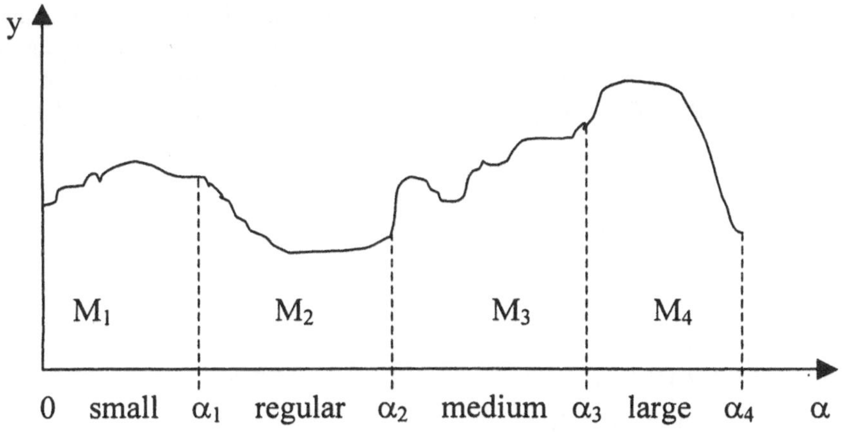

Figure 13.1 Modelling a complex dynamical system with the fuzzy inference system.

13.4 Adaptive Model-Based Control of International Trade

We describe in this section the application of our new method for adaptive model-based control to the problem of controlling international trade dynamics. The problem of international trade between three or more countries is a very complex one because of the couplings and non-linearities involved in the mathematical models (Castillo & Melin, 1998). In this section, we describe the methodology to develop an intelligent system for controlling international trade that can be used by the government of a specific country to maximize the profit from its international trade with other countries.

13.4.1 Adaptive model-based control of international trade

The method for adaptive model-based control of non-linear dynamical systems consists of using a fuzzy rule base for model selection, a genetic algorithm for identification, and a neural network for control. For the case of international trade, we need to define each of the method's components, mentioned above, to achieve the goal of controlling the dynamical system of three (or more) countries with trade between them.

The mathematical models of international trade can be represented as systems of coupled non-linear differential equations (as described in Section 13.2). In this case, we can establish a fuzzy rule base for model selection that enables the use of the appropriate mathematical model according to the changing conditions of the economies involved. For example, if we use the general mathematical models of Equations (13.3) and (13.4) for describing the international trade dynamics between one, two or three countries, we can have the following specific models. For one country with no international trade we have:

$$M_1: \quad y'_1 = \alpha_1(I_1 - S_1) \qquad (13.8)$$
$$r'_1 = \beta_1(L_1 - M_1/p_1)$$

For two countries with no international trade:

$$M_2: \quad y'_i = \alpha_i(I_i - S_i) \qquad i = 1,2 \qquad (13.9)$$
$$r'_i = \beta_i(L_i - M_i/p_i)$$

For two countries with international trade:

$$M_3: \quad y'_i = \alpha_i(I_i - S_i + \gamma\,(EX_i - IM_i)) \qquad i = 1,2 \qquad (13.10)$$
$$r'_i = \beta_i(L_i - M_i/p_i)$$

For three countries with no international trade:

$$M_4: \quad y'_i = \alpha_i(I_i - S_i) \qquad i = 1,2,3 \qquad (13.11)$$
$$r'_i = \beta_i(L_i - M_i/p_i)$$

And for three countries with international trade:

$$M_5: \quad y'_i = \alpha_i(I_i - S_i + \gamma\,(EX_i - IM_i)) \qquad i = 1,2,3 \qquad (13.12)$$
$$r'_i = \beta_i(L_i - M_i/p_i)$$

where I_i, S_i, L_i, M_i, EX_i, IM_i, and p_i are defined as in Section 2 of this chapter. Now, using γ as a selection parameter we can establish the fuzzy rule base for model selection as in Table 13.2.

Table 13.2 Fuzzy rule base for model selection of international trade

IF		**THEN**
γ	Number of countries	Mathematical Model
	one	M_1
small	two	M_2
large	two	M_3
small	three	M_4
large	three	M_5

In Table 13.2 we are assuming that the selection parameter γ can have only two possible fuzzy values (small and large). The reasoning behind this is that when γ is small, we can use the model with no international trade and when γ is large we can use the model with international trade. We have to note here that the fuzzy rule base has to be developed according to the particular case that is being considered.

The integration of the fuzzy rule base for model selection with the genetic algorithm for identification, and the neural network for control, results in an intelligent system for adaptive model-based control of international trade. This intelligent system combines the advantages of neural networks (ability for identification and control) with the advantages of fuzzy logic (use of expert knowledge), and the advantages of genetic algorithms (for optimization) to achieve the goal of robust adaptive control of international trade. The general architecture of the intelligent control system for international trade is similar to the one shown in (Melin and Castillo, 1998), except that now instead of the plant we have a non-linear dynamical system in economics. An intelligent system with this architecture is capable of adapting to changing conditions in the economies of the countries, because it can change the control actions according to the data available and also can change the reference mathematical model if there is a large enough change in the parameter γ. Of course, for this method to work we need to estimate parameter γ from time series of the real values for the variables in the mathematical models.

13.5 Simulation Results for Control of International Trade

To give an idea of the performance of our neuro-fuzzy-genetic approach for adaptive model-based control of international trade dynamics, we show below simulation results obtained for the case of three countries (USA, Canada and Mexico) with international trade. We will consider the problem of controlling the economy of the less developed country (Mexico) because it is the most challenging from the control point of view. For the case of Mexico, one problem is that of reducing interest rates in the short term so we will consider as a desired trajectory for this economy:

$$r_d = 0.25e^{-0.1t} + 0.02\sin t + 0.05$$

with initial values of :

$$r(0) = 0.30 \qquad r'(0) = 0.$$

In this desired trajectory for the economy, we are assuming that the goal interest rate is 5% and that we need to decrease the initial rate of 30% to the final interest rate of 5%. We also consider that the economy has natural cycles and because of this fact we use the "sine' function.

We use three-layer neural networks (with 10 hidden neurons) with the backpropagation algorithm and hyperbolic tangent sigmoidal functions as the activation functions for the neurons. We show in Figure 13.2 the initial function approximation achieved with the neural networks for control.

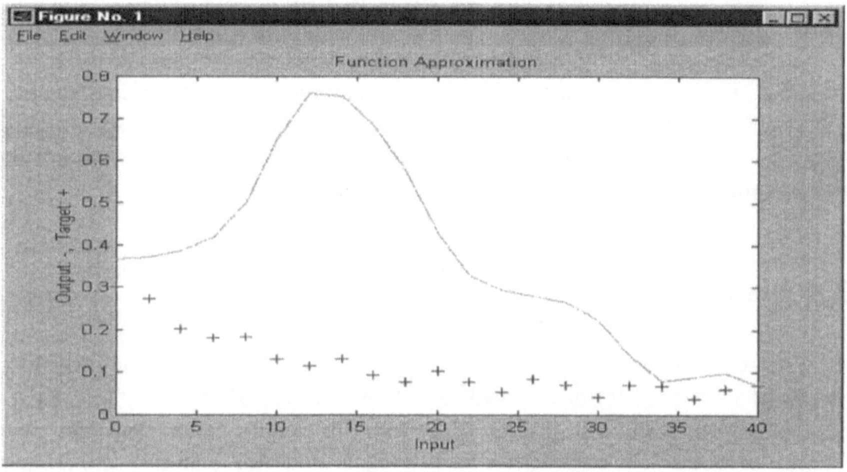

Figure 13.2 Initial function approximation of the neural network for control.

We show in Figure 13.3 the function approximation achieved with the neural network for control after 59 epochs of training with a variable learning rate. The identification achieved by the neural network (after 59 epochs) can be considered very good because the error has been decreased to the order of 10^{-4}. Still, we can obtain a better approximation by using more hidden neurons or more layers. In any case, we can see clearly how the neural network learns to control the economic dynamic system, because it is able to follow the arbitrary desired trajectory.

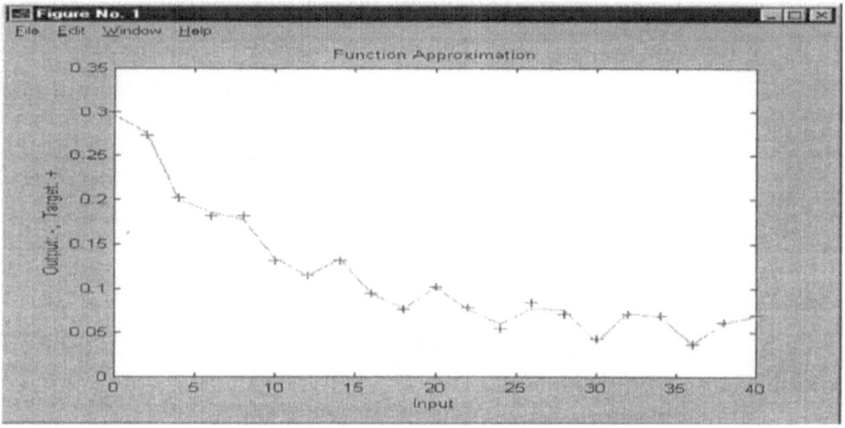

Figure 13.3 Function approximation of the neural network for control after 59 epochs.

We have to mention here that these simulation experiments for the case of three specific countries (USA, Canada and Mexico) show very good results. We have also tried our approach for control with other dynamic systems in economics with encouraging results. We recommend to the interested reader to use our methodology for this type of economic systems or other similar systems to explore on his (or her) own the interesting problem of controlling complex non-linear dynamical systems.

13.6 Summary

We described in this chapter a hybrid neuro-fuzzy-genetic approach for controlling international trade dynamics. The integration of the fuzzy rule base for model selection, with the genetic algorithm for identification and the neural network for control, results in an intelligent system for adaptive model-based control of international trade. This intelligent system was written in the MATLAB programming language (Nakamura, 1997) and combines the advantages of fuzzy logic (use of expert knowledge), the advantages of genetic algorithms (function optimization) and the advantages of neural networks (learning and adaptability) to achieve the goal of robust adaptive control of international trade. An intelligent system with this neuro-fuzzy-genetic approach is capable of adapting to changing conditions in the economies of the countries, because it can change the control actions according to the data available and also can change the reference mathematical model if there are large enough variations in the economies. We think that this hybrid neuro-fuzzy-genetic approach can be used for similar problems in economics and finance. Of course, in this case we will need to change the variables involved in the process of control.

References

Abraham, E. & Firth, W. J. (1984). "Multiparameter Universal Route to Chaos in a Fabry-Perot Resonator", Optical Bistability, Vol. 2, pp. 119-126.

Albertos, P., Strietzel, R. & Mart, N. (1997). "Control Engineering Solutions: A Practical Approach", IEEE Computer Society Press.

Badiru, A.B. (1992). "Expert Systems Applications in Engineering and Manufacturing", Prentice-Hall.

Barto, A. G., Sutton, R. S. & Anderson, C. (1983). "Neuronlike Elements that can Solve Difficult Learning Control Problems", IEEE Transactions on Systems, Man & Cybernetics, Vol. 13, pp. 835-846.

Bernard, J. A. (1988). "Use of Rule-Based System for Process Control", IEEE Control Systems Magazine, Vol. 8, pp. 3-13.

Bezdek, J.C. (1981). "Pattern Recognition with Fuzzy Objective Function Algorithms", Plenum Press, New York.

Bode, H., Brodd, R.J. and Kordesch, K.V. (1977). Lead-Acid Batteries, John Wiley & Sons.

Bratko, I. (1990). "Prolog Programming for Artificial Intelligence", Addison Wesley.

Brogan, W. (1991). "Modern Control Theory", Prentice-Hall.

Bryson, A. E. & Ho Y.-C. (1969). "Applied Optimal Control", Blaisdell Press.

Castillo, O. & Melin, P. (1994a). "Developing a New Method for the Identification of Microorganisms for the Food Industry using the Fractal Dimension, Journal of Fractals, World Scientific, Vol. 2, No. 3, pp. 457-460.

Castillo, O. & Melin, P. (1994b). "An Intelligent System for Discovering Mathematical Models for Financial Time Series Prediction", Proceedings of TENCON'94, IEEE Computer Society Press, Vol. 1, pp. 217-221.

Castillo, O. & Melin, P. (1995a). "An Intelligent System for Financial Time Series Prediction Combining Dynamical Systems Theory, Fractal Theory and Statistical Methods", Proceedings of CIFER'95, IEEE Computer Society Press, pp.151-155.

Castillo, O. & Melin, P. (1995b). "Intelligent Model Discovery for Financial Time Series Prediction using Non-Linear Dynamical Systems and Statistical Methods", Proceedings of the Third International Conference on Artificial Intelligence Applications on Wall Street", Software Engineering Press, pp. 80-89.

Castillo, O. & Melin, P. (1995c). "An Intelligent System for the Simulation of Non-Linear Dynamical Economical Systems", Journal of Mathematical Modelling and Simulation in Systems Analysis, Edited by Achim Sydow, Gordon and Breach Publishers, Vol. 18-19, pp. 767-770.

Castillo, O. & Melin, P. (1996a). "Automated Mathematical Modelling for Financial Time Series Prediction using Fuzzy Logic, Dynamical Systems and Fractal Theory", Proceedings of CIFER'96, IEEE Computer Society Press, pp. 120-126.

Castillo, O. & Melin, P. (1996b). "Automated Mathematical Modelling and Simulation of Dynamical Engineering Systems using Artificial Intelligence Techniques", Proceedings CESA'96, Gerf EC Lille, pp. 682-687.

Castillo, O. & Melin, P. (1996c). "An Intelligent System for Financial Time Series Prediction using Fuzzy Logic Techniques and Fractal Theory", Proceedings ITHURS'96, Vol. 1, AMSE Press, pp. 423-430.

Castillo, O. & Melin, P. (1997a). "Mathematical Modelling and Simulation of Robotic Dynamic Systems using an Intelligent Tutoring System based on Fuzzy Logic and Fractal Theory", Proceedings of AIENG'97, Wessex Institute of Technology, pp. 97-100.

Castillo, O. & Melin, P. (1997b). "Mathematical Modelling and Simulation of Robotic Dynamic Systems using Fuzzy Logic Techniques and Fractal Theory", Proceedings of IMACS World Congress'97, Wissenschaft & Technik Verlag, Vol. 5, pp.343-348.

Castillo, O. & Melin, P. (1998a). "A New Fuzzy-Fractal-Genetic Method for Automated Mathematical Modelling and Simulation of Robotic Dynamic Systems", Proceedings of World Congress on Computational Intelligence FUZZ'98, IEEE Computer Society Press, Vol. 2, pp. 1182-1187.

Castillo, O. & Melin, P. (1998b). "Modelling, Simulation and Behavior Identification of Non-Linear Dynamical Systems with a New Fuzzy-Fractal-Genetic Approach", Proceedings of IPMU'98, EDK Publishers, Vol. 1, pp. 467-474.

Castillo, O. & Melin, P. (1998c). "Modelling, Simulation and Forecasting of International Trade Dynamics using a New Fuzzy-Genetic Approach", Proceedings of CCM'98, AMSE Press, pp. 21-24.

Castillo, O. and Melin, P. (1999). "A New Fuzzy Inference System for Reasoning with Multiple Differential Equations for Modelling Complex Dynamical Systems", CIMCA'99, IOS Press, Vienna Austria, pp.224-229.

Castillo, O. and Melin, P. (1999). "Intelligent Model-Based Adaptive Control of Robotic Dynamic Systems with a New Neuro-Fuzzy-Genetic Approach", Proceedings of Robotics and Applications, Acta Press, Santa Barbara, USA. pp. 270-275.

Castillo, O. & Melin, P. (1999). "A General Method for Automated Simulation of Non-Linear Dynamical Systems using a New Fuzzy-Fractal-Genetic Approach", Proceedings CEC'99, IEEE Press, Vol. 3, pp. 2333-2340.

Castillo, O. & Melin, P. (1999). "Automated Mathematical Modelling, Simulation and Behavior Identification of Robotic Dynamic Systems using a new Fuzzy-Fractal-Genetic Approach", Journal of Robotics and Autonomous Systems, Vol. 28, No. 1, pp. 19-30.

Chalmers, D. (1990). " The evolution of learning: an experiment in genetic connectionism". Proceedings of the 1990 Connectionist Models Summer School. Morgan Kaufmann.

Chen, V. C. & Pao, Y. H. (1989). "Learning Control with Neural Networks", Proceedings of the International Conference on Robotics and Automation, pp. 1448-1453.

Chiu, S., Chand, S., Moore, D. & Chaudhary, A. (1991). "Fuzzy Logic for Control of Roll and Moment for a Flexible Wing Aircraft", IEEE Control Systems Magazine, Vol. 11, pp. 42-48.

Chua, L.O. (1993). "Global unfolding of Chua's circuit", IEICE Transactions Fundam., pp. 704-734.

Cybenko, G. (1989). "Approximation by Superpositions of a Sigmoidal Function", Mathematics of Control, Signals and Systems, Vol. 2, pp. 303-314.

Davidor, Y. (1991). "Genetic Algorithms and Robotics: A Heuristic Strategy for Optimization", World Scientific Publishing.

Devaney, R. (1989). "An Introduction to Chaotic Dynamical Systems", Addison Wesley Publishing.

Fahlman, S. E. & Lebiere C. (1990). "The Cascade-Correlation Learning Architecture", Advances in Neural Information Processing Systems, Morgan Kaufmann.

Fu, K.S., Gonzalez, R.C. & Lee, C.S.G (1987). "Robotics: Control, Sensing, Vision and Intelligence", McGraw-Hill.

Geman, S. & Geman, D. (1984). "Stochastic Relaxation, Gibbs Distribution and the Bayesian Restoration in Images", IEEE Transactions of Pattern Analysis and Machine Intelligence, Vol. 6, pp. 721-741.

Goldberg, D.E. (1989). "Genetic Algorithms in Search, Optimization and Machine Learning", Addison Wesley Publishing.

Grebogi, C., Ott, E. & Yorke, J. A. (1987). "Chaos, Strange Attractors, and Fractal Basin Boundaries in Nonlinear Dynamics", Science, Vol. 238, pp. 632-637.

Gujarati, D. (1987). "Basic Econometrics", McGraw-Hill Publishing.

Gupta, M. M. & Sinha, N. K. (1996). "Intelligent Control Systems: Theory and Applications", IEEE Computer Society Press.

Hanselman D. & Littlefield B. (1995). "The Student Edition of MATLAB Version 4 : User's Guide", The Math-Works, Inc. Prentice-Hall.

Hehner, N. and Orsino, J.A. (1985). Storage Battery Manufacturing Manual III, Independent Battery Manufacturers Association.

Holland, J. H. (1975). "Adaptation in Natural and Artificial Systems", University of Michigan Press.

Hunt, K. J., Sbarbaro, D., Zbikowski R. & Gawthrop, P. J. (1992). "Neural Networks for Control Systems-A survey", Automatica, Vol. 28 No. 6, pp. 1083-1112.

Ingber, L. & Rosen, B.E. (1992). "Genetic Algorithms and Very Fast Simulated Reannealing", Journal of Mathematical and Computer Modelling, Vol. 16, pp. 87-100.

Jamshidi, M. (1997). "Large-Scale Systems: Modelling, Control and Fuzzy Logic", Prentice-Hall

Jang, J.-S. R. (1993). "ANFIS: Adaptive-Network-Based Fuzzy Inference Systems", IEEE Transactions on Systems, Man and Cybernetics", Vol. 23, pp. 665-685.

Jang, J.-S. R. & Gulley, N. (1997). "MATLAB: Fuzzy Logic Toolbox, User's Guide", The Math-Works, Inc.Publisher.

Jang, J.-S. R., Sun, C.-T. & Mizutani, E. (1997). "Neurofuzzy and Soft Computing: A Computational Approach to Learning and Machine Intelligence", Prentice-Hall.

Kandel, A. (1992). "Fuzzy Expert Systems", CRC Press Inc.

Kapitaniak, T. (1996). "Controlling Chaos: Theoretical and Practical Methods in Non-Linear Dynamics", Academic Press.

Karnik, N.N & Mendel, J.M. (1998). "An Introduction to Type-2 Fuzzy Logic Systems", Technical Report, University of Southern California.

Kasai, Y. & Morimoto, Y (1988). "Electronically Controlled Continuously Variable Transmission", Proceedings of International Congress Transportation Electronics.

Kirkpatrick, S., Gelatt, C. D. & Vecchi, M. P. (1983). "Optimization by Simulated Annealing", Science, Vol. 220, pp. 671-680.

Kitano, H. (1990). "Designing neural networks using genetic algorithms with graph generation system", Journal of Complex Systems, Vol. 4, pp. 461-476.

Kocarev, L. & Kapitaniak, T. (1995). "On an equivalence of chaotic attractors", Journal of Physics A., Vol. 28, pp. 249-254.

Korn, G. A. (1995). "Neural Networks and Fuzzy Logic Control on Personal Computers and Workstations", MIT Press.

Kosko, B. (1992). "Neural Networks and Fuzzy Systems: A Dynamical Systems Approach to Machine Intelligence", Prentice-Hall.

Kosko, B. (1997). "Fuzzy Engineering", Prentice-Hall.

Lilly, K. W. (1993). "Efficient Dynamic Simulation of Robotic Mechanisms", Kluwer Academic Press.

Lim, S.Y., Hu, J. & Dawson, D.M. (1996). "An Output Feedback Controller for Trajectory Tracking of RLED Robots using Observed Backstepping Approach", Journal of Robotics and Automation, pp. 149-160.

Lippmann, R. P. (1987). "An Introduction to Computing with Neural Networks", IEEE Acoustics, Speech, and Signal Processing Magazine, Vol. 4, pp. 4-22.

Madan, R. (1993). "Chua's Circuit: Paradigm for Chaos", World Scientific, Singapore.

Mamdani, E. H. & Assilian, S. (1975). "An Experiment in Linguistic Synthesis with a Fuzzy Logic Controller", International Journal of Man-Machine Studies, Vol. 7, pp. 1-13.

Mandelbrot, B. (1987). "The Fractal Geometry of Nature", W. H. Freeman and Company.

Man, K. F., Tang, K.S. & Kwong, S. (1999). "Genetic Algorithms". Springer-Verlag.

Masters, T. (1993). "Practical Neural Network recipe in C++", Academic Press, Inc.

Melin, P. & Castillo, O. (1996). "Modelling and Simulation for Bacteria Growth Control in the Food Industry using Artificial Intelligence", Proceedings of CESA'96, Gerf EC Lille, pp. 676-681.

Melin, P. & Castillo, O. (1997a). " An Adaptive Model-Based Neural Network Controller for Biochemical Reactors in the Food Industry", Proceedings of Control'97, Acta Press, pp.147-150.

Melin, P. & Castillo, O. (1997b). "Mathematical Modelling and Simulation of Bacteria Growth in the Time and Space Domains using Artificial Intelligence, Dynamical Systems and Fractal Theory", Proceedings of AMS'97, Acta Press, pp. 484-487.

Melin, P. & Castillo, O. (1997c). "An Adaptive Neural Network System for Bacteria Growth Control in the Food Industry using Mathematical Modelling and Simulation", Proceedings of IMACS World Congress'97, Wissenschaft & Technik Verlag, Vol 4 pp. 203-208.

Melin, P. & Castillo, O. (1997d). "Automated Mathematical Modelling and Simulation for Bacteria Growth Control in the Food Industry using Artificial Intelligence and Fractal Theory", Journal of Systems Analysis, Modelling and Simulation, Edited by Achim Sydow, Gordon and Breach Publishers, Vol. 29, pp. 189-206.

Melin, P. & Castillo, O. (1998a). "An Adaptive Model-Based Neuro-Fuzzy-Fractal Controller for Biochemical Reactors in the Food Industry", Proceedings of IJCNN'98, IEEE Computer Society Press, Vol. 1, pp.106-111.

Melin, P. & Castillo, O. (1998b). "A New Method for Adaptive Model-Based Neuro-Fuzzy-Fractal Control of Non-Linear Dynamic Plants: The Case of Biochemical Reactors", Proceedings of IPMU'98, EDK Publishers, Vol. 1, pp. 475-482.

Melin, P. and Castillo, O. (1999). A New Method for Adaptive Model-Based Neuro-Fuzzy-Fractal Control of Non-Linear Dynamical Systems, Proceedings of the International Conference of Non-Linear Problems in Aviation and Aerospace'98, European Conference Publications, Daytona Beach, USA, pp. 499-506

Melin, P. and Castillo, O. (1999). "A New Neuro-Fuzzy-Fractal Approach for Adaptive Model-Based Control of Non-Linear Dynamic Plants", ", Proceedings of Intelligent Systems and Control, Acta Press, Santa Barbara, USA. pp. 397-401.

Miller, G., Todd,P. & Hedge, S. (1989). "Designing Neural Networks using Genetic Algorithms", Proceedings of the Third International Conference on Genetic Algorithms", Morgan Kauffmann.

Miller, W. T., Sutton, R. S. & Werbos P. J. (1995). "Neural Networks for Control", MIT Press.

Minsky, M. & Papert, S. (1969). "Preceptrons", MIT Press.

Mitchell, M. (1996). "An Introduction to Genetic Algorithms", MIT Press.

Montana, G. & Davis, L. (1989). "Training feedforward networks using genetic algorithms", Proceedings of the International Joint Conference on Artificial Intelligence, Morgan Kauffmann.

Morari, M. & Zafiriou, E. (1989). "Robust Process Control", Prentice-Hall.

Moulet, M. (1992). "A Symbolic Algorithm for Computing Coefficients Accuracy in Regression", Proceedings of the International Workshop on Machine Learning, pp. 332-337.

Nakamura, S. (1997). "Numerical Analysis and Graphic Visualization with MATLAB", Prentice Hall.

Narendra, K. S. & Annaswamy, A. M. (1989). "Stable Adaptive Systems", Prentice Hall Publishing Company.

Ng, G. W. (1997). "Application of Neural Networks to Adaptive Control of Non-Linear Systems", John Wiley & Sons.

Ogorzalek, M. J. (1993). "Taming Chaos-Part II: Control. IEEE Transactions on Circuit Systems, Vol. 40, pp. 700-721.

Omidvar, O. & Elliot, D. L. (1997). "Neural Systems for Control", Academic Press.

Ott, E., Grebogi, C. & Yorke, J.A. (1990). "Controlling Chaos". Physical Review Letters, Vol. 64, pp. 1196-1199.

Otten, R. H. J. M. & van Ginneken, L. P. (1989). "The Annealing Algorithm", Kluwer Academic.

Pelczar, M. J. & Reid, R. D. (1982). "Microbiology", McGraw Hill.

Parker D. B. (1982). "Learning Logic", Invention Report S81-64, File 1, Office of Technology Licencing.

Pham, D. T. & Xing, L. (1995). "Neural Networks for Identification, Prediction and Control", Springer-Verlag.

Pomerleau, D. A. (1991). "Efficient Training of Artificial Neural Networks for Autonomous Navigation", Journal of Neural Computation, Vol. 3, pp. 88-97.

Psaltis, D., Sideris, A. & Yamamura, A. (1988). "A Multilayered Neural Network Controller", IEEE Control Systems Magazine, Vol. 8, pp.17-21.

Pyragas, K. (1992). "Continuous control of chaos by self-controlling feedback", Physical Letters, pp. 421-428.

Rao, R. B. & Lu, S. (1993). "A Knowledge-Based Equation Discovery System for Engineering Domains", IEEE Expert, pp. 37-42.

Rasband, S.N. (1990). "Chaotic Dynamics of Non-Linear Systems", Wiley Interscience.

Rich, E. & Knight, K. (1991). "Artificial Intelligence", McGraw-Hill.

Rosenblatt, F. (1962). "Principles of Neurodynamics: Perceptrons and the Theory of Brain Mechanisms", Spartan.

Ruelle, D. (1990). "Deterministic Chaos: The Science and the Fiction", Proc. Roy. Soc. London, Vol. 427, pp. 241-248.

Rumelhart, D. E., Hinton, G. E. & Williams, R. J. (1986). "Learning Internal Representations by Error Propagation", Parallel Distributed Processing: Explorations in the Microestructure of Cognition, Vol. 1, Chapter 8, pp. 318-362, MIT Press.

Runkler, T. A. & Glesner, M. (1994). "Defuzzification and Ranking in the Context of Membership Value Semantics, Rule Modality, and Measurement Theory", Proceedings of European Congress on Fuzzy and Intelligent Technologies.

Russell, S. & Norvig, P. (1995). "Artificial Intelligence: A Modern Approach", Prentice-Hall.

Sackinger, E., Boser, B. E., Bromley, J., LeCun, Y. & Jackel, L. D. (1992). "Application of the Anna Neural Network Chip to High-Speed Character Recognition", IEEE Transactions on Neural Networks, Vol. 3, pp. 498-505.

Samad, T. & Foslien, W. (1994). "Neural Networks as Generic Nonlinear Controllers", Proceedings of the World Congress on Neural Networks, pp. 191-194.

Sejnowski, T. J. & Rosenberg, C. R. (1987). "Parallel Networks that Learn to Pronounce English Text", Journal of Complex Systems, Vol. 1, pp. 145-168.

Sepulveda, R., Castillo, O., Montiel, O. and Lopez, M. (1998). "Analysis of Fuzzy Control System for the Process of Forming Batteries", Proceedings of ISRA'98, Coahuila, Mexico, pp. 203-210.

Sleeman, D. & Edwards, P. (1992). "Proceedings of the International Workshop on Machine Learning", Morgan Kauffman Publishers.

Soucek, B. (1991). "Neural and Intelligent Systems Integration: Fifth and Sixth Generation Integrated Reasoning Information Systems", John Wiley and Sons.

Staib, W. E. (1993). "The Intelligent Arc Furance: Neural Networks Revolutionize Steelmaking", Proceedings of the World Congress on Neural Networks, pp. 466-469.

Staib, W. E. & Staib, R. B. (1992). "The Intelligent Arc Furnace Controller: A Neural Network Electrode Position Optimization System for the Electric Arc Furnace", Proceedings of the International Conference on Neural Networks, Vol. 3, pp. 1-9.

Su, H. T. & McAvoy, T. J. (1993). "Neural Model Predictive Models of Nonlinear Chemical Processes", Proceedings of the 8th International Symposium on Intelligent Control, pp. 358-363.

Su, H. T., McAvoy, T. J. & Werbos, P. J. (1992). "Long-term Predictions of Chemical Processes using Recurrent Neural Networks: A Parallel Training Approach", Industrial & Engineering Chemistry Research, Vol. 31, pp. 1338-1352.

Sueda, N. & Iwamasa, M. (1995). "A Pilot System for Plant Control using Model-Based Reasoning", IEEE Expert, Vol. 10, No. 4, pp.24-31.

Sugeno, M. & Kang, G. T. (1988). "Structure Identification of Fuzzy Model", Journal of Fuzzy Sets and Systems", Vol. 28, pp. 15-33.

Szu, H. & Hartley, R. (1987). "Fast Simulated Annealing", Physics Letters, Vol. 122, pp. 157-162.

Takagi, T. & Sugeno, M. (1985). "Fuzzy Identification of Systems and its Applications to Modeling and Control", IEEE Transactions on Systems, Man and Cybernetics, Vol. 15, pp. 116-132.

Troudet, T. (1991). "Towards Practical Design using Neural Computation", Proceedings of the International Conference on Neural Networks, Vol. 2, pp. 675-681.

Tsukamoto, Y. (1979). "An Approach to Fuzzy Reasoning Method", In Gupta, M. M., Ragade, R. K. and Yager, R. R., editors, Advanced in Fuzzy Set Theory and Applications, pp. 137-149, North-Holland.

Ungar, L. H. (1995). "A Bioreactor Benchmark for Adaptive Network-Based Process Control", Neural Networks for Control, MIT Press, pp. 387-402.

Von Altrock, C. (1995). "Fuzzy Logic & Neuro Fuzzy Applications Explained", Prentice Hall.

Wagenknecht, M. & Hartmann, K. (1988). "Application of Fuzzy Sets of Type 2 to the solution of Fuzzy Equation Systems", Fuzzy Sets and Systems, Vol. 25, pp. 183-190.

Weigend, A. & Gershenfeld, N.A. (1994). "Time Series Prediction: Forecasting the Future and Understanding the Past", Addison Wesley Publishing.

Werbos, P. J. (1991). "An Overview of Neural Networks for Control", IEEE Control Systems Magazine, Vol. 11, pp. 40-41.

Werbos, P. J. (1974). "Beyond Regression: New Tools for Prediction and Analysis in the Behavioral Sciences", Ph.D. Thesis, Harvard University.

Widrow, B. & Stearns, D. (1985). "Adaptive Signal Processing", Prentice-Hall.

Yager, R.R. (1980). "Fuzzy Subsets of Type II in Decisions", Journal of Cybernetics, Vol. 10, pp. 137-159.

Yager, R.R. & Filev, D.P. (1993). "SLIDE: A Simple Adaptive Defuzzification Method", IEEE Transactions on Fuzzy Systems, Vol. 1, pp. 69-78.

Yager, R. R. & Filev, D.P. (1994). "Generation of Fuzzy Rules by Mountain Clustering", Journal of Intelligent and Fuzzy Systems", Vol. 2, No. 3, pp. 209-219.

Yamamoto, Y. & Yun, X. (1997). "A Modular Approach to Dynamic Modelling of a Class of Mobil Manipulators", Journal of Robotics and Automation. pp. 41-48.

Yasunobu, S. & Miyamoto, S. (1985). "Automatic Train Operation by Predictive Fuzzy Control", Industrial Applications of Fuzzy Control, pp.1-18, North Holland.

Yen, J., Langari, R. & Zadeh, L. A. (1995). "Industrial Applications of Fuzzy Control and Intelligent Systems", IEEE Computer Society Press.

Zadeh, L.A. (1965). "Fuzzy Sets", Journal of Information and Control, Vol. 8, pp. 338-353.

Zadeh, L.A. (1971). "Similarity Relations and Fuzzy Ordering", Journal of Information Sciences", Vol. 3, pp. 177-206.

Zadeh, L.A. (1973). "Outline of a New Approach to the Analysis of Complex Systems and Decision Processes", IEEE Transactions on Systems, Man and Cybernetics, Vol. 3, pp. 28-44.

Zadeh, L.A. (1975). "The Concept of a Linguistic Variable and its Application to Aproxímate Reasoning –1", Information Sciences, Vol. 8, pp. 199-249.

Zomaya, A. Y. (1992). "Modelling and Simulation of Robot Manipulators: A Parallel Processing Approach", World Scientific Publishing.

Index